AF412951

Molecules as Components of Electronic Devices

ACS SYMPOSIUM SERIES **844**

Molecules as Components of Electronic Devises

Marya Lieberman, Editor
University of Notre Dame

American Chemical Society, Washington, DC

Library of Congress Cataloging-in-Publication Data

Molecules as components of electronic devises / Marya Lieberman, editor.

p. cm.—(ACS symposium series ; 844)

Includes bibliographical references and index.

ISBN 0–8412–3782–4

1. Molecular electronics.

I. Lieberman, Marya, 1967-. II. Series.

TK7874.8 .M69   2003
621.381—dc21   2002028144

Foreword

The ACS Symposium Series was first published in 1974 to provide a mechanism for publishing symposia quickly in book form. The purpose of the series is to publish timely, comprehensive books developed from ACS sponsored symposia based on current scientific research. Occasionally, books are developed from symposia sponsored by other organizations when the topic is of keen interest to the chemistry audience.

Before agreeing to publish a book, the proposed table of contents is reviewed for appropriate and comprehensive coverage and for interest to the audience. Some papers may be excluded to better focus the book; others may be added to provide comprehensiveness. When appropriate, overview or introductory chapters are added. Drafts of chapters are peer-reviewed prior to final acceptance or rejection, and manuscripts are prepared in camera-ready format.

As a rule, only original research papers and original review papers are included in the volumes. Verbatim reproductions of previously published papers are not accepted.

ACS Books Department

Contents

Indexes

Preface

During the past half century, electronic devices have gotten smaller, lighter, and cheaper at the same time their capabilities have increased. The integrated circuit is an excellent case in point; the miniaturization of integrated circuit elements has been a reliable trend in the past 40 years. However, even the most optimistic observer cannot extrapolate this trend forever. The 2001 International Technology Roadmap for Semiconductors states, "at 10–15 years in the future, it becomes evident that most of the known technological capabilities will approach or have reached their limits."[1]

Integrated circuits are assembled from a very limited palette of materials. The materials used in most integrated circuits are single-crystal silicon, a few types of dopant atom, silicon dioxide or nitride, and aluminum or copper. Today, researchers and engineers are trying to add more materials to the palette by using molecules as components of electronic devices. Molecules can do things that solid-state devices cannot, such as recognize other molecules or self-assemble into simple structures. In theory, molecules can do many of the same things that solid-state devices can, but in a much smaller area of real estate on a chip or with much smaller power consumption or cost. A few successful examples, such as liquid crystalline materials used in display screens, have filled niches in the device world for decades and will be familiar to anyone reading this preface. Others, like memory chips made from molecules, seem more the stuff of science fiction.[2]

Until fairly recently, it was difficult to cut single molecules out of the herd in order to study their properties, and unclear what properties would lead to useful devices or how to organize the molecules so they could interact in a controlled way with the macroscopic world. With the

---

[1] International Technology Roadmap for Semiconductors, 2001 Edition

[2] "Memory," Lois McMaster Bujold, Baen; 1996

Figure 1. Comparison of a 1 kilobyte magnetic core memory, in use in the early 1960s (the large circuit board) and a more recent 1 megabyte memory chip introduced in the mid-1980s (mounted on a 1" dot at the top of the circuit board). Modern proposals for molecular electronic memories are based on the same crossbar architecture that is used in the magnetic core memory, but their proposed integration density would approach 1 terabyte/cm2.

development of techniques that can measure or manipulate single molecules, such as scanning tunneling spectroscopy, these questions are finally finding answers. In turn, measurements of molecular properties have inspired speculation and new experiments on how those properties might be turned to advantage in order to accomplish electronic functions.

Work on molecular computing received a big boost from Defense Advanced Research Projects Agency in the late 1990s. A large injection of funds to academic researchers and government labs produced new collaborations and a burst of significant improvements in fabrication, characterization, measurement, and theoretical understanding of molecular devices for transmitting, storing, and processing information. *Science* magazine ("Molecules Get Wired," Robert F. Service, *Science* **Dec. 21, 2001**,2442–2443) selected molecular electronics as the scientific advance of the year for 2001, a year which also saw the completion of the draft human genome. Companies also stepped up the pace of research in molecular electronics. Several large electronics companies (HP, IBM, Bell Labs) have set up research divisions to work on varying aspects of molecular electronics, and dozens of start-up companies are working to manufacture products in which molecules serve electronic functions. Cheap organic LEDs, ultra-compact flash memories, programmable labels on the supermarket shelves, and hybrid organic–inorganic photovoltaic devices are examples of what may be coming down the pike.

This book is based on a symposium on molecular electronics held at the Spring 2001 American Chemical Society National meeting in San Diego, California. The symposium was organized to bring together researchers who work on different aspects of molecular electronics. The contributors range from synthetic organic and inorganic chemists to physical chemists. Contributors submitted recent research reports in three areas: measurements, materials, and theory. The first section, Measurements, deals with the electrical conductivity and charge retention properties of self-assembled monolayers of molecules. These properties are relevant to molecules that function as insulators, diodes, memory elements, or two-terminal transistors. Wide-area measurements made on metal–SAM–metal devices and scanning-probe measurements of properties of single molecules are included. The section on Materials is more wide ranging in scope; it includes work on nanoparticle and device fabrication for photovoltaic devices and sensors, and the synthesis and testing of specialized molecules for functions such as light emission. The Theory section includes ab initio studies of electron transport through molecules and an introduction to molecular photovoltaic devices. The introductory chapter describes a graduate level course in molecular electronics, with specific resources for instructors who want to tackle this topic.

Marya Lieberman

Department of Chemistry and Biochemistry
University of Notre Dame
South Bend, IN 46556
574–631–4665 (telephone)
574–631–6652 (fax)
mlieberm@nd.edu (email)

Chapter 1

A Crash Course in Molecular Electronics

M. Lieberman

Department of Chemistry, University of Notre Dame,
South Bend, IN 46556 (mlieberm@nd.edu)

A short, discussion-based course in molecular electronics for graduate and advanced undergraduate students was developed and tested at the University of Notre Dame in spring 2002. A short set of lectures conveys necessary background information. After this, students lead 8-10 sessions in which important papers from the recent molecular electronics literature are critically discussed.

Introduction

With recent publicity directed at molecular electronics and more research groups venturing into the field, there are many students who are curious about molecular electronics. This chapter describes a crash course in molecular electronics that is based on student-led discussion of primary papers from the recent literature. The course is suitable for a semester-long seminar series or a 4-week segment of a regular MWF graduate course. The first segment of the course consists of basic information and references for five core background areas; this information can be presented effectively by an instructor using a lecture format. The bulk of the course content, though, consists of critical discussion of primary papers in the field. Guidelines are given for helping students lead these discussions, and a suggested list of papers on molecular electronics is given. They include some classics in this young field (eg theoretical work by Aviram and Ratner on molecular rectifiers, Adleman's DNA computation paper, the Reed/Tour break junction work) and more recent work on molecular-scale components for logic and memory (architectural papers, QCA, SETs, and cross-bar devices). There are certainly many other papers on topics such as device fabrication, computing architectures, conducting polymers, and light emitting or photovoltaic devices, which would fit under the heading of molecular electronics, and which the reader should feel free to include if he or she wishes.

Making the discussion format work

Students should think critically about the methods, results, and conclusions of the papers they read. One pedagogical technique which elicits the necessary mental effort is to use a discussion format and make the students responsible for leading the discussions. The students who are not assigned to lead the discussion on a particular paper tend to read the paper superficially and accept its conclusions at face value; this tendency can be counteracted by a well-designed set of reading questions. The discussion leader constructs the reading questions for a particular paper before he or she leads the discussion of that paper. The instructor must review these questions before the discussion. The review provides a natural check point for the instructor to make sure that the student has a) read the paper, b) grasped the important issues without misconceptions, and c) made an effort to impose some intellectual structure on their new knowledge.

The instructor's role consists of reviewing the discussion questions with the student, discussing goals and strategies for the class session that the student is to lead, refraining from taking over the class discussion, and conducting a "post-mortem" after the class to help the student understand what parts of the discussion went well and what could use polishing. The amount of time required for this process is about the same as that required to prepare a lecture on the material, but the students seem to learn a lot more from the discussion format than from a lecture. In particular, students are much more willing to admit ignorance and to answer each other's questions, so the discussion format elicits many more interactions among the students and reflective comments than a lecture.

Useful pedagogical guidelines for the student discussion leaders, which should be gone over in a one-on-one setting before the discussion and revisited at the post-mortem after the discussion:

- The discussion leader's role is to guide the discussion and help the whole class advance their understanding of the paper, not to give a lecture or literature seminar.
- Use the discussion questions to structure the discussion. Figure out ahead of time how long you want to spend on each question.
- Do not pose a question and then answer it yourself. Silence is painful, but the discussion is advanced only when the other students participate. Instead, try rephrasing the question or breaking it down into simpler parts.
- One goal should be to elicit comments and questions from everyone in the class. Focus the group's attention on each other's

contributions by using people's names when you refer to something they said, e.g., "Peng, what did you think of Amy's interpretation of Figure 1c?" Don't let one person hog the discussion. If someone has not participated in the discussion, call on them by name to answer a simple question or comment on another student's statement. If someone asks you a question, it's fine to toss it back to the class.

- Summarize the results of the discussion on one topic or question before you move on to the next.

- For students who are not native English speakers: Go over the paper and your discussion questions carefully with the instructor ahead of time to make sure you understand the paper, have phrased your questions comprehensibly, and know how to pronounce the terms in the paper.

Background

The core material needed to understand the papers will depend on the nature of the papers chosen and the backgrounds of the students. The following topics can be covered at a superficial level in about five lectures, and will provide enough background for most chemistry students to understand the molecular electronics papers listed in the next section.

a) Fabrication: Any microchip fabrication text[1] will have good resources for understanding photolithography, and a range of more specialized techniques for sub-200 nm fabrication has been reviewed by Wallruff and Hinsberg.[2] The synthesis and characterization of colloidal metal particles was recently reviewed by Bawendi et al.[3] Carbon nanotubes[4] and semiconductor or metal nanowires[5] are also important components of cross-bar devices.

b) Review of electronic circuitry: An excellent review of symbols, units, series vs. parallel behavior and I-V curves for circuit elements such as diodes and resistors can be garnered from Horowitz and Hill's classic electronics text.[6]

c) Oriented arrays of molecules: Ulman's 1996 review[7] offers a complete overview of the thiol-on-gold and siloxane-on-oxide SAMs used in many molecular electronic devices; his earlier book[8] goes into more detail on Langmuir-Blodget films and gives an excellent overview of the characterization techniques which are used to determine the structure and order of LB films and SAMs.

d) Electrochemistry: Any general chemistry text[9] will contain a serviceable refresher on the redox properties of molecules. J. Chem. Ed. has accessible articles on electron transfer at electrodes[10] and cyclic voltammetry.[11] Lindsey et

4

al. discuss the issues involved in making electrical contacts to molecular monolayers.[12] The relationship between molecular orbitals and vacuum energy levels is discussed in some depth in a recent paper by Tian *et al.*[13]

e) DNA and PCR (needed for the Adleman DNA computing paper): Any molecular biology text[14] will contain a discussion of DNA structure, base pairing, melting, mismatches and thermodynamics, primers and enzymes, and gel electrophoresis. The technique of PCR is explained in layman's language[15] and in more technical detail[16] in fairly recent articles.

Acknowledgement:

Thanks to Prof. Peter Kogge in the Notre Dame Dept of Electrical Engineering, whose discussion group on "Revolutionary Ideas in Computing" was the model for this course, and to C. Fennell, D. Fogarty, W. He, J. Jiao, P. Sun, S. Taylor, A. Vickers, and S. Vijay, who served as guinea pigs in Spring 2002.

Suggested papers and discussion guidelines

In each case, one or more references are given as well as guidelines for the student discussion questions.

1) Molecular Rectifiers, A. Aviram and M.A. Ratner, Chem. Phys. Lett, *29,* 277-283, **1974** Alternatively, one could use the review by R. M. Metzger (J. Mater. Chem., **1999**, *9,* 2017-2036)

For Ratner's paper, it is crucial that everyone understand what a rectifier is and what its I-V curve should look like. Make sure the students understand how the parts of the molecules in Figures 1 & 2 correspond to the energy level diagram in Fig 3, and that they understand how electrons and holes move under positive and negative bias. It is worth it to spend a fair amount of time on Figs. 4 and 6.

If the Metzger review is discussed in the same class session as the Ratner paper, it's best to focus on one system (eg starting on p. 2031) and go through the fabrication, characterization, and electrical measurements in detail. Metzger discusses several alternate interpretations of the data and this discussion should be followed closely.

2) Molecular Computation of Solutions to Combinatorial Problems, L. M. Adleman, Science, *266,* 1021-1024, **1994**. (see also letters in Science *268,* 481-484, **1995**)

Students must have a working understanding of PCR in order to make sense of this paper. The key to a good discussion is mapping the algorithm on p. 1021 to the experimental manipulations of PCR. It might be of interest to discuss the letters in light of recent DNA computation work done on surfaces.[17]

3) Conductance of a Molecular Junction, M. A. Reed, C. Zhou, J. Muller, T. P. Burgin, J. M. Tour, Science *278*, 252-253, **1997**

A good starting point is the schematic in Fig. 1; make sure that all parts of the device are understood. Students may enjoy coming up with alternatives to the scheme in Fig. 3. The core of the paper is the experimental measurements shown in fig. 4, check for understanding of conductance vs. current and make sure students think about why a "gap" is observed and how it might relate to the properties of the molecule in the junction.

4) Self-assembly of single electron transistors and related devices, D. L. Feldheim and C. D. Keating, Chem. Soc. Revs. *27* 1-12 **1998**

This is a fairly hefty paper but can be discussed in one class period. Everyone needs to be clear on the idea of Coulomb blockade (e.g. figure 4) and a review of capacitor properties may be useful. Focus on Section 3: pick one or more systems and go through fabrication, characterization, and measurement. If there is time, move on to section 4 and discuss applications. Students should be able to suggest some potential fabrication challenges.

5) Papers on architecture: a) Device architecture for computing with quantum dots, C. S. Lent and P. D. Tougaw, Proc. IEEE *85*, 541-557, **1997**, b) A Defect-Tolerant Computer Architecture: Opportunities for Nanotechnology, J. R. Heath, P. J. Kuekes, G. S. Snider, and R. S. Williams, Science, *280*, 1716-1721, **1998**, c) Architectures for Molecular Electronic Computers. 1. Logic Structures and an Adder Built from Molecular Electronic Diodes (the "pink book"). James C. Ellenbogen and J. Christopher Love, Nanosystems Group, The MITRE Corporation, **1999**, can be downloaded from http://www.mitre.org/technology/nanotech/Arch_for_MolecElec_Comp_1.html (accessed 7 May 2002)

Molecular electronics will require some overall architectural scheme in order to produce useful devices. These three papers could each be the topic of a class session--don't try to discuss more than one in a session. For architecture papers, the discussion should focus on understanding the how the properties of the basic molecular device are incorporated into a larger scheme. Make sure everyone is clear on a) what the molecular device does and b) how they interact with each other. Architectures should be able to handle errors in fabrication, and thermodynamic factors may be important. It is useful to discuss implementation at a practical level--how is I/O handled, what are the main challenges for fabricating the architecture, how does the architecture differ from other proposals.

6) Observation of Switching in a quantum-dot cellular automata cell, G. H. Bernstein, I. Amlani, A. O. Orlov, G. S. Lent, and G. L. Snider, Nanotechnology, *10*, 166-173, **1999**

If paper 5b has not been discussed, a brief overview of QCA architecture would be a good starting point. If it has, then start by asking about the difference between the cell shown in Fig. 1 and the ones discussed by Lent and Tougaw. Figure 3 should be discussed in detail to make sure everyone understands how the two schematic representations of the cell are related to the physical implementation. Two methods are used for probing the charge state of the cell, conductance and SET electrometer, these should both be discussed. Students may want to discuss problems for larger-scale implementation if there is time.

7) Electronically Configurable Molecular-Based Logic Gates, C. P. Collier; E. W. Wong, M. Belohradsky, F. M. Raymo, J. F. Stoddart, P. J. Kuekes, R. S. Williams, and J. R. Heath, Science *285*, 391-394, **1999**

It is helpful to have the students try to draw these devices; a lot of them have trouble understanding what the actual device consists of and how it is related to the schematics in Fig. 1. Make sure they understand how the rotaxane is incorporated into the device. The idea of resonant tunneling is a crucial one, and should be discussed along with Fig. 1c. Next, focus on the experimental data in Fig. 3; the goal here should be to understand what is happening to the molecule as the electrode Fermi levels are changed. Figure 4 requires that the students understand both how the basic device operates, and the path of current through the crossbar arrays shown in schematic form. It's worth the time for the students to draw the current path explicitly.

8) Logic gates and computation from assembled nanowire building blocks, Y. Huang; X. F. Duan; Y. Cui, L. J. Lauhon; K. H. Kim; and C. M. Lieber, Science *294*, 1313-1317, **2001** Logic Circuits with Carbon Nanotube Transistors, A. Bachtold, P. Hadley, T. Nakanishi, and C. Dekker, Science Science *294*, 1317-1320, **2001**

Both these papers rely on a background understanding of pn junctions and FETs. A quick review would be a good starting point for the discussion.

For the Lieber paper: start with fabrication (how are nanowires doped and assembled) and proceed to measurement of I-V properties, then discuss Figure 2, the logic gates. Here it is necessary to understand how the physical implementation relates to the circuit diagrams and to step through at least one of the truth tables.

The Dekker paper is closely related to the Lieber paper; it's not necessary to do both papers together, but it does work in a single class session. Again, start with fabrication and then discuss measurement of the electrical properties of the crossed nanotubes. Make sure students understand where the gate voltage is applied and its effect on conductivity. If there is time, pick one or two of the devices in Figure 4 to discuss.

References

[1] For example, "Microchip Fabrication," P. Van Zant, 4th Ed. McGraw-Hill, NY, 2000, chapter 4.

[2] "Lithographic Imaging Techniques for the Formation of Nanoscale Features," Wallruff, G. M. and Hinsberg, W. D.; Chem. Rev. (1999) 99(7) 1801-1821.

[3] "Synthesis and characterization of monodisperse nanocrystals and close-packed nanocrystal assemblies," Murray CB, Kagan CR, Bawendi MG, Ann. Rev. Mater. Sci., (2000) 30 545-610

[4] "Nanotubes from Carbon," Ajayan, P. M. (1999) 99(7) 1787-1800

[5] a) "Synthetic control of the diameter and length of single crystal semiconductor nanowires," Gudiksen, M. S.; Wang, J.; and Lieber, C. M. J. Phys. Chem. B (2001), 105(19) 4062-4064. b) " Template synthesis of metal nanowires containing monolayer molecular junctions," Mbindyo JKN, Mallouk TE, Mattzela JB, et al. JACS (2002) 124(15) 4020-4026

[6] "The Art of Electronics, 2nd Edition," Paul Horowitz and Winfield Hill; Cambridge, Cambridge University Press: 1989. See Ch. 1 and first half of Ch. 2.

[7] "Formation and structure of self-assembled monolayers," A. Ulman, Chem Rev. (1996) *96* 1533-1554.

[8] "An introduction to ultrathin organic films : from Langmuir-Blodgett to self-assembly," Ulman, Abraham; Boston : Academic Press, 1991.

[9] For example, "Principles of Modern Chemistry," 4th Ed., D. W. Oxtoby, H. P. Gillis, and N. H. Nachtrieb, Brooks/Cole 2002, pp. 174-176 and Ch. 12.

[10] "Understanding electrochemistry: some general concepts," Larry R. Faulkner, J. Chem. Ed.1983, 60, 262-264

[11] "An Introduction to Cyclic Voltammetry," Gary A. Mabbott, J. Chem. Ed., 1983, 60, 697-701.

[12] "Making electrical contacts to molecular monolayers," XD, Zarate X, Tomfohr J, Sankey OF, Primak A, Moore AL, Moore TA, Gust D, Harris G, Lindsay SM, Nanotechnology (2002) 13(1) 5-14.

[13] " Current-Voltage Characteristics of Self-Assembled Monolayers by Scanning Tunneling Microscopy," S. Datta, W. Tian, S. Hong, R. Reifenberger, J. I. Henderson, and C. P. Kubiak, Phys. Rev. Lett. (1997) 79, 2530–2533

[14] See for example "Molecular Cell Biology, 3rd Edition," H. Lodish, D. Baltimore, A. Berk, S.L. Zipursky, P. Matsudaira, and J. Darnell, New York: Scientific American Books: 1995, pp 101-119 and Ch 7. Also of interest:

"Programmed Materials Synthesis with DNA," Storhoff, J. J. and Mirkin, C. A., Chem. Rev. (1999) 99(7) 1849-1862.

[15] "PCR," Mullis, K. B., Sci. Am. (1990) 262(4) p. 56-65.

[16] "Specific Synthesis of DNA *in Vitro* via a Polymerase-Catalyzed Chain Reaction," Mullis, K. B. and Faloona, F. Methods Enzymol. (1987) 155 335-350.

[17] "DNA computing on surfaces," Liu QH, Wang LM, Frutos AG, Condon AE, Corn RM, Smith LM, Nature (2000) 403 (6766): 175-179.

Measurements

Using Probe Lithography and Self-Assembled Monolayers To Investigate Potential Molecular Electronics Systems

R. Lloyd Carroll, Ryan Fuierer, and Christopher B. Gorman[*]

Department of Chemistry, North Carolina State University, Raleigh, NC 27695

The pursuit of Molecular Electronics – that is, molecules that can be used as transistors, switches, rectifiers, and even logic gates – is a large and growing field of study. A fundamental phenomenon displayed by many essential molecular electronics components is that of Negative Differential Resistance (NDR). It is expected that molecules with accessible redox states (i.e., electroactive species) should display NDR. Scanning Probe Lithography was used to fabricate isolated nanostructures of redox-active molecules, and these were observed to display strong NDR under ambient conditions[1].

The implementation of Molecular Electronics has two general requirements: the identification and preparation of molecular species with well-defined electronic properties and the ability to position, find, and interact with the molecules. Scanning Probe Lithography offers a method of prototyping molecular candidates, using the tip as both pencil and probe. Scanning Probe Lithography is useful to position small numbers of molecules with precision, find them in a background of inert molecules, and test their electronic properties while in a well-defined, surface bound nanostructure. Probe Lithography techniques utilizing both Scanning Tunneling Microscopy (STM) and Atomic Force Microscopy (AFM) have been demonstrated.[2-8] Our work has focused on the implementation of in-situ replacement lithography utilizing the STM[9], with imaging and lithography under an inert organic fluid such as dodecane. There are several benefits to such an in-situ process. Foremost, the organic fluid provides a method of mass transport, in that interesting molecules can be dissolved in the fluid and replaced into an existent surface. Moreover, inert organic fluids of very low polarity minimize leakage currents between the tip and sample, allowing imaging at very low tunneling currents.

The process of *in situ* replacement lithography is straightforward. A self-assembled monolayer (or SAM) of an alkylthiolate (typically formed from decanethiol or dodecanethiol) is prepared on a freshly annealed Au(111) surface. This SAM provides a well-ordered, largely crystalline surface into which other functionalized molecules can be replaced. An approximately 1 mM solution of a second functional thiol in dodecane is added before imaging. Imaging the SAM (typically at 1 V and 5-10 pA tunneling current) under the solution does not effect replacement of the second thiol. Raising the bias between tip and sample above some threshold voltage (typically ~3 V) promotes replacement. Setting this bias and moving the tip in some computer-controlled pattern causes desorption of the thiolate SAM beneath the tip, and allows adsorption of the solvated thiol into the vacancies created (Figure 1). Upon completion of the pattern, the bias is again lowered to allow imaging of the surface without causing replacement. Line resolution using this technique is 10-15 nm[9].

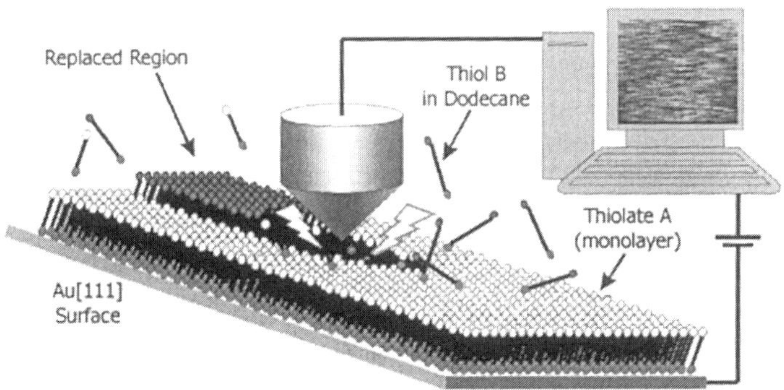

Figure 1. Schematic of replacement step in Scanning Probe Lithography

Figure 2 illustrates the characteristics of a ferrocene-terminated undecanethiolate SAM (henceforth "**Fc**") formed by replacement into an inert background of dodecanethiolate SAM. At low bias, the replaced **Fc** SAM is largely indistinguishable from the background. However, upon increasing the bias, the **F c** SAM appears to increase in height, compared to the dodecanethiolate. This apparent height contrast indicates an enhanced flow of current through the **Fc** molecules, compared to the dodecanethiolate.

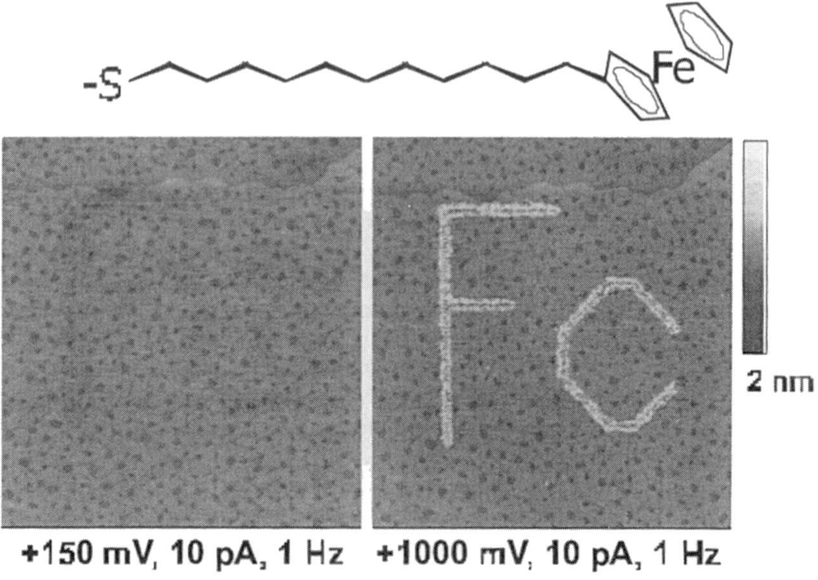

+150 mV, 10 pA, 1 Hz   +1000 mV, 10 pA, 1 Hz

*Figure 2. Bias Dependent Contrast exhibited by **Fc** replaced into dodecanethiolate SAM. STM images: 350nm scan size, indicated imaging parameters, 2nm z-scale, under dodecane*

To investigate the origin of the bias dependent contrast, 1024 I-V curves were collected over a patterned surface composed of a square of **Fc** replaced into a background of decanethiolate (Figure 3, inset), under dodecane. The I-V curves were collected by turning off the feedback between tip and surface, then sweeping from positive bias to negative bias, in a specified range. The current response was collected and stored. The graph in Figure 3 shows two I-V curves. The solid line is a typical curve selected from within the replaced **Fc** region. The non-linearity in the I-V curve is a phenomenon called Negative Differential Resistance, or NDR. The actual mechanism giving rise to the NDR in this system has not been established with certainty, but presumably, it arises due to resonant tunneling through accessible redox states in the **Fc** molecules. This is a similar phenomenon to that found in resonant tunneling diodes (RTDs).[13] The dashed line is a typical curve selected from the decanethiolate region, outside the replacement square. Since the decanethiolate does not contain low-lying, accessible redox states, resonant tunneling in this bias range was not expected. Indeed, no NDR was observed. No apparent NDR behavior was observed when other, non-electroactive SAMs (e.g., those composed of carboxylic acid terminated alkylthiols) were probed. Similar NDR behavior was observed in both air and under dodecane solution.

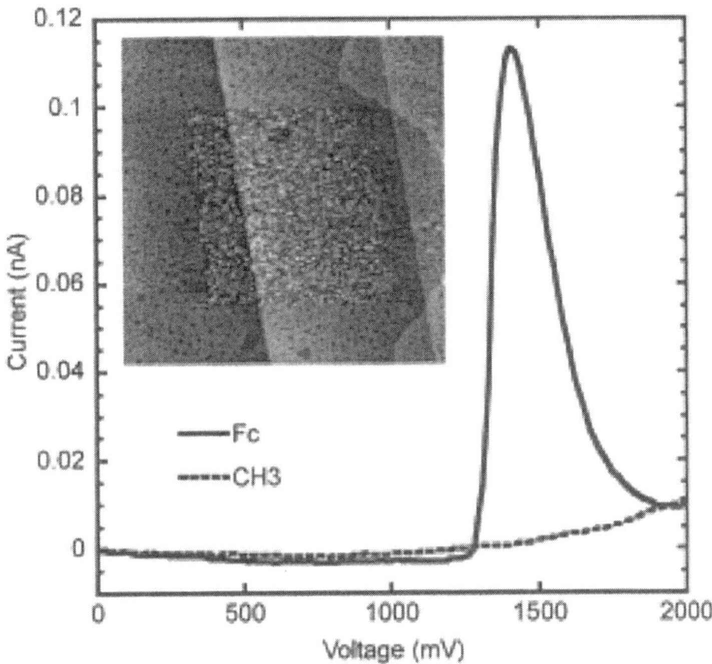

*Figure 3. NDR in **Fc** SAM. The inset is an STM image (700nm scan size, with a 400nm replaced region, collected at 1V, 5pA, 1Hz scan rate, under dodecane) of a replaced region of **Fc** SAM in a background of decanethiolate SAM. The two I-V curves are representative curves from the two regions of the surface: the solid line is collected over the **Fc** SAM, and the dashed line is collected over decanethiolate SAM. (Adapted from reference 1. Copyright 2001 American Chemical Society.)*

From the I-V curves of **Fc**, it is apparent that the bias dependent contrast exhibited by **Fc** during imaging at higher biases arises as a consequence of the difference in current flow through the two types of molecules. Over **Fc**, as the tip-sample bias approaches the position of NDR, enhanced current flow through the **Fc** molecule leads to retraction of the tip, which is displayed as a larger apparent height contrast between **Fc** and decanethiolate.

Upon examination of the I-V curves collected in many experiments, it was observed that the position at which the NDR occurs shifts slightly from curve to curve. This shift in NDR has been observed by others[10-12]. Given that the I-V curves are collected in a two electrode configuration (that is, tip and substrate), there is no chemical reference, and it is expected that the NDR position should shift somewhat during an experiment. To determine the variability of the response, a homogeneous **Fc** SAM was prepared and I-V curves collected over the SAM, under dodecane. A histogram of the observed peak position in a data set of 1015 curves is shown in Figure 4. A Gaussian fit of the histogram gives a mean peak position of 1.5 V with a full-width at half maximum of 0.1 V.

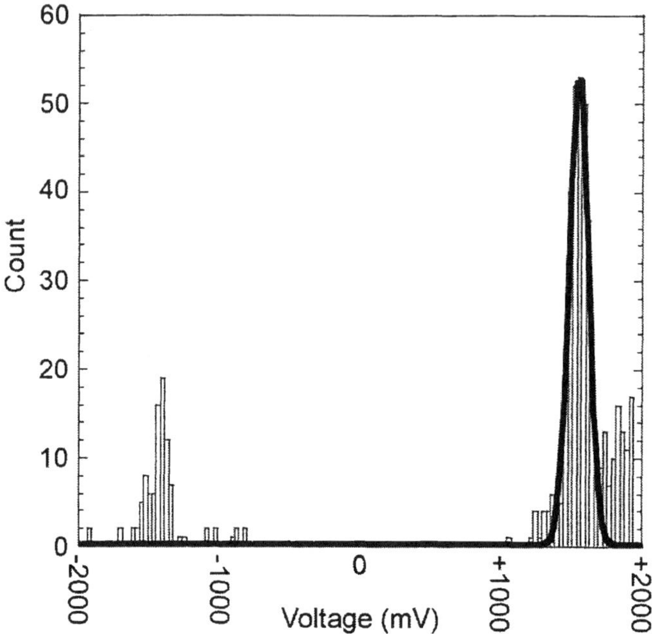

*Figure 4. Histogram of position of NDR in an **Fc** SAM, under dodecane. (Adapted from reference 1. Copyright 2001 American Chemical Society.)*

It was hypothesized that other redox-active molecules would also display molecular NDR. Galvinol-terminated hexanethioacetate ("**Gal**", Figure 5) was provided by Professor David Shultz, and used in a "two-ink" experiment in which first **Fc**-SAM and then subsequently **Gal**-SAM regions were defined with replacement lithography into a dodecanethiolate SAM. As shown in the two panels of Figure 5 collected at low and high biases under dodecane, both molecules exhibit bias dependent contrast, within similar ranges. Further work is underway to synthesize other redox-active species, with particular effort towards those with accessible redox states that are different from **Fc** and **Gal**.

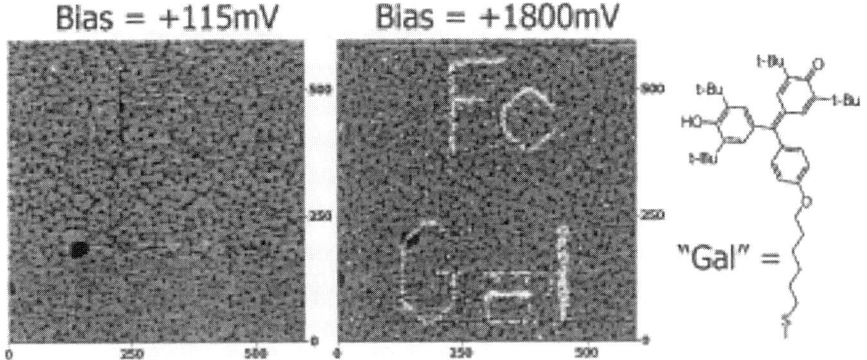

*Figure 5. Bias dependent contrast in **Fc** and **Gal** in a "two-ink" experiment. STM images: 600x600nm, indicated bias, 10pA, 1Hz scan rate. (Adapted from reference 1. Copyright 2001 American Chemical Society)*

Acknowledgments

This work was supported in part by the Air Force Office of Scientific Research MURI Program in Nanoscale Chemistry, and by the National Science Foundation (CAREER Award, DMR-9600138 and CHE-9900072).

References

(1) Gorman, C. B.; Carroll, R. L.; Fuierer, R. R. *Langmuir* **2001**, *17(22)*, 6923-6930.

(2) Schoer, J. K.; Zamborini, F. P.; Crooks, R. M. *J. Phys. Chem.* **1996**, *100*, 11086-11091.

(3) Schoer, J. K.; Crooks, R. M. *Langmuir* **1997**, *13*, 2323-2332.

(4) Xu, S.; Liu, G. Y. *Langmuir* **1997**, *13*, 127-129.

(5) Mizutani, W.; Ishida, T.; Tokumoto, H. *Langmuir* **1998**, *14*, 7197-7202.

(6) Chen, J.; Reed, M. A.; Asplund, C. L.; Cassell, A. M.; Myrick, M. L.; Rawlett, A. M.; Tour, J. M.; Van Patten, P. G. *Appl. Phys. Lett.* **1999**, *75*, 624-626.

(7) Piner, R. D.; Zhu, J.; Xu, F.; Hong, S. H.; Mirkin, C. A. *Science* **1999**, *283*, 661-663.

(8) Xu, S.; Miller, S.; Laibinis, P. E.; Liu, G. Y. *Langmuir* **1999**, *15*, 7244-7251.

(9) Gorman, C. B.; Carroll, R. L.; He, Y. F.; Tian, F.; Fuierer, R. *Langmuir* **2000**, *16*, 6312-6316.

(10) Chen, J.; Reed, M. A.; Rawlett, A. M.; Tour, J. M. *Science* **1999**, *286*, 1550-1552.

(11) Kinne, M.; Barteau, M. A. *Surf. Sci.* **2000**, *447*, 105-111.

(12) Han, W. H.; Durantini, E. N.; Moore, T. A.; Moore, A. L.; Gust, D.; Rez, P.; Leatherman, G.; Seely, G. R.; Tao, N. J.; Lindsay, S. M. *J. Phys. Chem. B* **1997**, *101*, 10719-10725.

(13) Sze, S. M. *Physics of Semiconductor Devices*; 2nd ed.; John Wiley & Sons: New York, 1981.

Chapter 3

Molecular Electronics with a Metal–Insulator–Metal Junction Based on Self-Assembled Monolayers

Michael L. Chabinyc[1], R. Erik Holmlin[1], Rainer Haag[1], Xiaoxi Chen[1], Rustem F. Ismagilov[1], Maria A. Rampi[2,*], and George M. Whitesides[1,*]

[1]Department of Chemistry and Chemical Biology, Harvard University, 12 Oxford Street, Cambridge, MA 02138
[2]Dipartimento di Chimica, Centro di Fotochimica, CNR, Università di Ferrara, 44100 Ferrara, Italy

The mechanisms of electron transport in metal-insulator-metal junctions are incompletely understood. A metal-insulator-metal junction consisting of a self-assembled monolayer (SAM) supported on a mercury drop in mechanical contact with a SAM on a planar metal electrode has been developed as a test-bed with which to study electron transport through organic films. This review provides a summary of results intended to characterize this junction including: i) the determination of the electrical breakdown field of organic monolayers, ii) the determination of the tunneling decay constant for aliphatic and aromatic organic oligomers, and iii) the examination of molecular rectifier.

Introduction

An objective in molecular (or organic) electronics is to understand electron transport in organic-metallic heterostructures. The achievement of this goal will require experimental methods that allow broad structural classes of organic materials to be studied, and that yield results that are unambiguously interpretable. This review describes the development of a new metal-insulator-metal (MIM) junction. The junction is composed of a self-assembled monolayer (SAM) supported on a mercury drop that is placed in conformal mechanical contact with a SAM on a metal electrode. It is relatively simple to fabricate and requires little capital investment. One of the main benefits of this junction is its convenience for the study of electron transport through thin organic films.

Why molecular electronics?

The use of organic and organometallic molecules as constituent elements of electronic devices has become relevant for several reasons. Within the last decade (1990's), the feature sizes of the components of semiconductor-based electronics (~ 150 nm) have begun to approach the molecular scale.[1] At molecular length scales (< 10 nm), quantum behavior may provide the basis for new types of electronic devices.[2,3] The cost of building new semiconductor fabrication facilities has also grown enormously in the last twenty years; if an electronic device could be developed that would be assembled under conditions that did not require stringent environmental control, the cost to fabricate the device would certainly be reduced. These issues, and others, have stimulated efforts to determine if molecular species are plausible candidates for active components in electronic devices.

Organic and organometallic molecules are attractive alternatives to solid-state materials at the nm-scale because of their electronic structure and their potential for self-assembly. Molecules inherently exhibit quantum–mechanical behavior at room temperature.[4,5] The electronic properties of molecules can be tailored readily by chemical synthesis; synthesis thus, in principle, provides a means to tune the behavior of molecular electronic devices. Self-assembly has emerged as a promising strategy for the fabrication of nm-scale structures by a bottom-up approach.[6-8] An ideal system would be one in which the molecules were pre-programmed by synthesis with the information required to assemble into a working device. Methods for the self-assembly of both molecules and nanoparticles of metals and semiconductors have been developed.[9-14] These methods may provide a means for the fabrication of heterostructures of organic molecules and solid-state materials.

18

What is required to make a functional molecular electronic device?

Building practical molecular electronic components requires solutions to a number of problems. The challenges include the development of new fabrication methods, and the development of predictive models of electron transport in molecular systems. These two issues are intertwined. The chemical functionalities that are used to direct assembly can affect the electronic properties of molecules[5]; for example, Reed has shown that the chemical group used to attach organic molecules to gold surfaces affects the height of the barrier for electron injection in a MIM structure.[15]

New fabrication methods that define both the structures of molecular devices and also the structures of their interconnects must be developed. A practical scheme to 'wire' molecular devices into computational structures using other molecules has yet to be developed. In addition, molecular devices must still have connections to external components for inputs and outputs. These external components may not be molecular in composition; we therefore need to explore methods for the integration of heterostructures.

Our current understanding of electron transport in nm–scale organic-metallic heterostructures is based on phenomenology. If we are truly to create working devices, we must develop a predictive physical-organic theory for the behavior of molecular systems. At this point, we do not know of a realistic architecture for a computational device that is entirely composed of molecular components.[5,16] The most widely applied approach to the fabrication of molecular electronic devices is to find molecular systems that mimic the function of devices used in the current generation of semiconductor microelectronic devices. Examples of these components include wires, resistors, capacitors, diodes, and transistors.[17] This approach may not prove to be the best ultimate strategy to achieve a molecular computer, but the information gained from studies of these standard components will certainly increase our ability to predict the characteristics of electron transport through organic matter at the nanometer scale.

Methods for the Study of Electron Transport in Organic Matter.

The ability of molecules to mediate electron transfer between chemical groups, or between electrodes and molecules, has been studied for many years.[18-24] The data from these studies have demonstrated that we understand, in part, how to predict the consequences of changes in molecular structure for the rates of electron transfer. Most understanding of electron transfer is based on measurements made on solution-based systems. The assumptions made in the predictive models for these systems do not necessarily hold for the same molecules in other environments (e.g. solid-state junctions). It is important to understand how the results on electron transfer obtained from a variety of experimental techniques are related.

Most studies of electron transfer in solution-based systems have been performed using two techniques – photoinduced electron transfer in donor-bridge-acceptor (D-B-A) systems[25,26] and electrochemical kinetics.[27,28] In experiments based on photoinduced electron transfer, the rate of electron transfer between an electron donor and electron acceptor has been measured as a function of the bridging species.[19,25,26] In electrochemical systems, the rate of electron transfer between a metal electrode that is covered by an organic film and a redox species in solution is typically measured.[27,29-33] These types of studies have proven most useful for uncovering the distance dependence of the rate of electron transfer through organic oligomers (Table I). The distance dependence for an oligomer, generally symbolized by "β" and expressed in the units Å^{-1}, follows an empirical exponential relation of the form, $k_{ET} = k_0 e^{-\beta d}$ (k_{ET} is the rate constant for electron transfer, d is the distance between the species that the electron is being transferred between, and k_0 is a preexponential factor). These studies demonstrate that organic oligomers facilitate electron transfer relative to the rate of electron tunneling through vacuum.

Table I. Values of β for Organic Molecules Measured with Different Experimental Systems.

System	*Composition of Organics*	*$\beta\,(\text{Å}^{-1})$*	*Ref*
D-B-A	Saturated Hydrocarbon	0.8-1.0	25, 26
	Oligophenylene	0.4-0.6	25, 26
Electrochemical	Alkanethiol SAM on Au	0.9-1.2	28
	Oligo(phenyleneacetylene)	0.4-0.5	30, 33
MIM Junction	SAM of Fatty Acid on Al/Al$_2$O$_3$	1.5	34
J$_{\text{Hg-SAM//SAM-Hg}}$ Junction	Alkanethiol SAM on Hg	0.8	38
CP-AFM	Alkanethiol SAM on Au	1.1	46, 47
STM	Alkanethiol SAM on Au	1.2	45

Metal-insulator-metal (MIM) junctions have been used to study transport through thin organic films.[4,34-38] Early work by Mann and Kuhn demonstrated that junctions with an insulating layer of a monolayer of organic material could be fabricated and tested.[34] A number of studies on MIM junctions have examined the distance dependence of the rate of tunneling.[34,38] Recently, a number of functional MIM junctions have been demonstrated. Metzger was the first to demonstrate unimolecular rectification conclusively in a MIM junction and has studied the electrical properties of these systems extensively.[37] Reed and Tour have shown that 100-nm^2 area MIM junctions that are fabricated with nitro-substituted aryl thiols exhibit negative differential resistance.[39] Heath, Stoddart, and co-workers have demonstrated MIM junctions based on Lagmuir-Blodgett films of electroactive rotaxanes that show switchable current-voltage relationships.[40] Both of these devices can be used as the basis of memory circuits. Although these MIM systems have demonstrated interesting functionalities, there are a number of experimental difficulties that complicate interpretation of these results. One of the most important of these difficulties is the nature of the interface formed by the evaporation of metal atoms onto an organic film. In addition, technical problems in the fabrication junctions without electrical shorts have limited the reproducibility of the results in some cases.[41]

Scanning probe microscopies (SPM) have been applied to the study of molecular systems.[42,43] These methods have the possibility of studying transport through single molecules.[44] In many cases, the interpretation of the results of STM experiments is difficult due to the convolution of the tip-substrate distance with the conductance. Despite this difficulty, Weiss and co-workers were able to measure the distance dependence of tunneling through alkanethiols.[45] Conducting probe atomic force microscopies (CP-AFM) provide a method to place the tip accurately at a fixed distance from the sample.[46] Frisbie has performed ensemble electrical measurements of SAMs using CP-AFM with a bare gold tip and with a gold tip coated by a SAM.[47] This work, and earlier work by Salmeron, demonstrated that the tunneling decay constant can be measured using MIM junctions implemented by CP-AFM.[46] SPM has not only demonstrated that electron transport can be studied, but also that these techniques provide new methods of controlling electron transport. Joachim and Gimzewski have shown that a single C_{60} molecule can be used as a transistor if the "gate" is provided by mechanical force placed on the molecule by a scanning probe tip.[48] A similar actuation scheme that uses electrostatic force has been demonstrated as a means to control transport in carbon nanotubes.[49] While the results from these studies are impressive, they require the use of specialized equipment.

We have begun to study a new system that is simple to assemble and is easily used by non-specialists. The system consist of a junction fabricated from a SAM supported by a mercury drop in mechanical contact with a SAM supported on a metal electrode.[50] We believe that this system will enable a wider range of molecular systems to be studied than has previously been possible.

A Metal/SAM//SAM/Mercury Junction

Fabrication.

The metal/SAM//SAM/mercury junction ("/" indicates a covalently bonded interface and "//" indicates a van der Waals interface) is formed by mechanical contact of a SAM supported on a mercury drop and a SAM supported on a metal electrode (Figure 1). To describe these junctions, we use the nomenclature $J_{Ag-SAM(1)//SAM(2)-Hg}$, where J indicates a junction, SAM(1) indicates the SAM on the solid Ag electrode, and SAM(2) indicates the SAM on the Hg electrode. The fabrication process of the metal/SAM//SAM/mercury junction is quite simple. First, SAM(1) is formed by exposure of a film of Au or Ag supported on a silicon wafer to a solution of a thiol or disulfide for ~12 hours. The SAM-coated electrode is placed in a hexadecane bath containing the thiol used for SAM(2). A 1-mL syringe that is filled with mercury is placed in the bath and a small drop of mercury is formed at the tip of the syringe. SAM(2) forms on the Hg drop electrode by absorption of the thiol from the hexadecane solution. The two electrodes are connected to the leads of an electrometer and are kept at ground potential. After ~5-10 min., the SAM-Hg droplet is brought into mechanical contact with the SAM-metal electrode using a micromanipulator. The micromanipulator also allows for lateral translation of the syringe and precisely controls the position of contact between the SAM-Hg drop and the SAM-metal electrode. The process is monitored using a CCD camera with a magnifying lens (20x). The area of contact is estimated by measuring the distance across the contact region of the mercury drop and solid electrode using the image from the CCD camera. The geometry of the contact region is assumed to be circular.

Measurement of Current-Voltage Curves.

The current-voltage (I-V) curves for a junction were measured using a Keithley 6517 electrometer as the source of bias potential and as the ammeter.[51] The potential ramp was applied to the junction as a series of small steps (typically 50 mV) with a ~5 s interval between them. The time interval and small voltage step-size ensured that the current was measured at a steady state. The current measured through the junction at a fixed bias voltage is stable over approximately one thousand seconds (Figure 2). The junction is stable to repeated potential cycling over hour-long periods. The current through a junction can drift upwards by as much as a factor of ~3 over an hour period during repeated measurements. Majda has reported substantially larger drifts (factor of > 10) in Hg/SAM//SAM/Hg junctions with SAMs formed from alkanethiols.[52] The origin of the difference in stability between the junctions describe here, which have one liquid and one solid electrode, and those described by Majda, which have two liquid electrodes, is unclear.

Electrical Measurements with Mercury/SAM//SAM/Metal Junctions

Electrical Breakdown Field of Organic SAMs.

The breakdown field of an insulator defines the operational range of voltages that can be applied to a MIM device. The breakdown field is defined as the value of the electrical field at which the conductance of a junction rapidly increases and the device fails irreversibly.[53] The nature of the change in a material during the breakdown process is not well understood for any system, organic or inorganic. Comparisons of the breakdown fields for different systems are difficult to make because the breakdown process is strongly affected by the presence of impurities. Knowledge of the breakdown fields of organic monolayers is especially important for molecular electronic devices because the fields that are created in thin-film junctions can be greater than 1 GV/m at a relatively low applied bias potentials (e.g. at a bias potential of 5 V, the field in a junction with an insulating layer of 2 nm is ~ 2 GV/m). We have used the metal/SAM//SAM/mercury junction to determine the limits of the value of the electric field that can be applied to SAM-based junctions.[54]

We have confirmed that the SAM-mercury drop conforms to the surface of the SAM on the opposite electrode. The values of capacitance were obtained for a series of junctions formed from alkanethiols. The measured values agree with values estimated using a parallel plate capacitor model parameterized with a thickness estimated from the experimental geometries.[55] The capacitances of junctions formed and characterized in ethanol and in hexadecane are similar within an order of magnitude. For example, a junction $J_{Ag-C16//C16-Hg}$ of area 0.4 mm^2 formed in ethanol has a capacitance of 1.5 nF; the same junction formed in hexadecane has a capacitance of 0.5 nF. Majda has also reported similar values for similar junctions with two mercury electrodes.[38,52] We do not know if the small difference between the values (a factor of ~3) implies that solvent molecules are intercalated in the junction or if a simple parallel plate capacitor

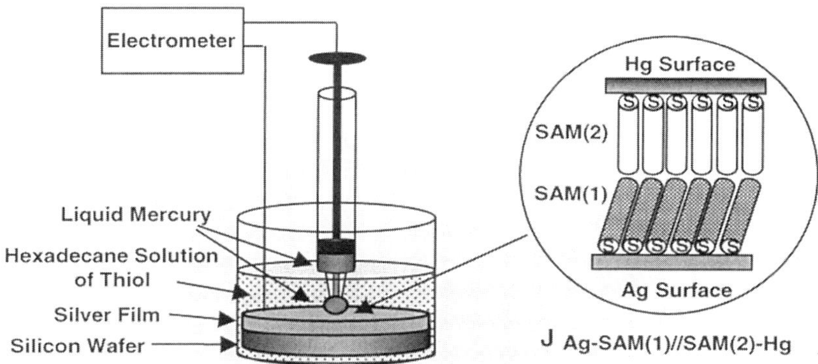

Figure 1. *Schematic illustration of a $J_{Ag\text{-}SAM(1)//SAM(2)\text{-}Hg}$ junction (the nomenclature is explained in the text).*

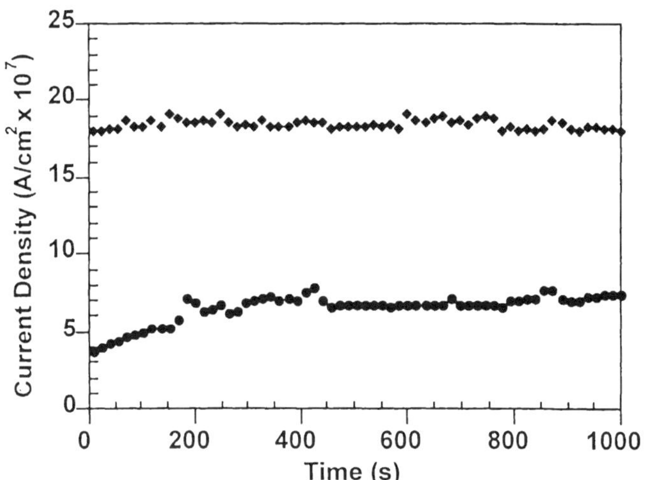

Figure 2. *Plot of current as a function of time for a junction with the structure $J_{Ag\text{-}C10//C16\text{-}Hg}$. The first set of data (•) was measured by increasing the bias on the junction from 0 to 0.5V over 40s. The bias was then held fixed at 0.5V (this time period is shown in the plot). The second set of data (♦) was taken using the same procedure on the same junction approximately 20 min. later. During the 20 min. period between these measurements the bias on the junction was held at 1V (data not shown).*

model that entirely neglects the medium in the region near the edges of the circumference of the contact region is insufficient to model the junction. Nonetheless, these data strongly suggest that conformal contact exists in the system.

We measured the breakdown fields of aliphatic and aromatic molecules (Table II).[54] The breakdown field for aliphatic SAMs (0.5 GV/m) is similar to that for polyethylene (0.6-0.8 GV/m). Aromatic SAMs exhibit breakdown fields in our junction that are similar to those of aliphatic compounds. Interestingly, Reed and co-workers were able to apply fields that are nearly one order of magnitude higher on nm^2–area MIM junctions formed with aromatic SAMs and upper electrodes formed by evaporation of metal on the organic layer.[56] We do not understand if this result is due to the fact these junctions are fabricated on, essentially, single grains of gold or if the chemical nature of the junction is affected by the deposition of the upper electrode.

This work demonstrates that organic films having thicknesses of > 2 nm can withstand the application of substantial electric fields (~ 0.5 GV/m). The effect of these fields on the electrical transport properties of molecules in MIM junctions has yet to be established; for example, the electronic levels of many molecular systems are known to shift at electrical fields of this magnitude.

Table II. Dielectric Strengths of Ag/SAM//SAM/Hg Junctions

SAM(1)	SAM(2)	Breakdown Voltage (V)	Field Strength (GV/m)
$HS(CH_2)_7CH_3$	$HS(CH_2)_{15}CH_3$	1.7 ± 0.4	0.5 ± 0.1
$HS(CH_2)_{11}CH_3$	$HS(CH_2)_{15}CH_3$	1.9 ± 0.1	0.5 ± 0.1
$HS(CH_2)_{15}CH_3$	$HS(CH_2)_{15}CH_3$	2.5 ± 0.4	0.5 ± 0.1
$HS(CH_2)_9CH_3$	$HS(CH_2)_{13}CH_3$	1.9 ± 0.2	0.5 ± 0.1
$HS(CH_2)_{11}CH_3$	$HS(CH_2)_{11}CH_3$	1.9 ± 0.3	0.5 ± 0.1
$HS\text{-}CH_2\text{-}C_6H_4\text{-}C_6H_5$	$HS(CH_2)_{15}CH_3$	1.3 ± 0.3	
$HS\text{-}C_6H_4\text{-}C_6H_5$	$HS(CH_2)_{15}CH_3$	0.7 ± 0.3	
polyethylene			0.6-0.8*
Teflon			25*
Pyrex			0.17*

SOURCE: All values from Reference 55. Those marked '*' from Reference 53.

Distance Dependence of Tunneling in Organic SAMs

Knowledge of the distance dependence of the tunneling current through organic thin films is necessary for the design of appropriate molecular linkers in molecular electronics. The rates of electron transfer through organic molecules have been measured using electrochemical and D–B–A systems (Table I). These data have shown that the dependence of the rate of electron transfer on distance follows an empirical exponential relation of the form, $k_{ET} = k_0 e^{-\beta d}$ (k_{ET} is the rate constant for electron transfer, d is the distance between the species that the electron is being transferred between, and k_0 is a preexponential factor). A similar exponential relationship for the distance dependence of electron transport was found by Mann and Kuhn using data from a MIM junction composed of a L-B film supported on an aluminum oxide layer with an evaporated upper electrode of aluminum.[34] In this case, the distance dependence of the current density is measured rather than a rate constant for electron transfer, i.e. $I(V) \propto e^{-\beta d}$ (I is the current density, d is the distance between the electrodes, and β is the distance dependence). Majda has recently used a Hg-SAM//SAM/Hg junction to determine the distance dependence of tunneling through MIM junctions.[38] These studies have demonstrated the utility of the distance dependence of electron transport as an metric to verify the quality of an organic MIM junction.

We studied the distance dependence of the magnitude of current flowing through a series of metal/SAM//SAM/mercury junctions[57,58] and used the data to test a new theoretical model for electron transport in molecular junctions.[59]

We established that the mechanism of electron transport in the junctions is tunneling.[58] The current-voltage (I-V) curves exhibit a nearly linear region at small bias voltage (< 0.1V) and a nonlinear region (Figure 3). This shape is expected for transport in a tunnel junction based on a simple model of a one-dimensional square barrier. The energetic separation of the HOMO and LUMO for an alkane chain is ~ 8 eV.[60] If the Fermi level of the electrode lies midway in this gap, the barrier height should be ~ 4 eV. Even at room temperature, this barrier height is large enough to ensure that transport occurs by electron tunneling.

We measured the I-V curves for a series of SAMs of aliphatic and aromatic oligomers (Figure 4). The distance dependence of the tunneling current, β, through aliphatic and aromatic oligomers was established from these data (Figure 5).[58] The value of β for aromatic oligomers (0.6 Å^{-1}) was smaller than that for aliphatic compounds(0.9 Å^{-1}). This result was expected based on the relative separation of the electronic energy levels in these molecules and agrees with previous data from scanning probe[45] and solution-based measurements.[30]

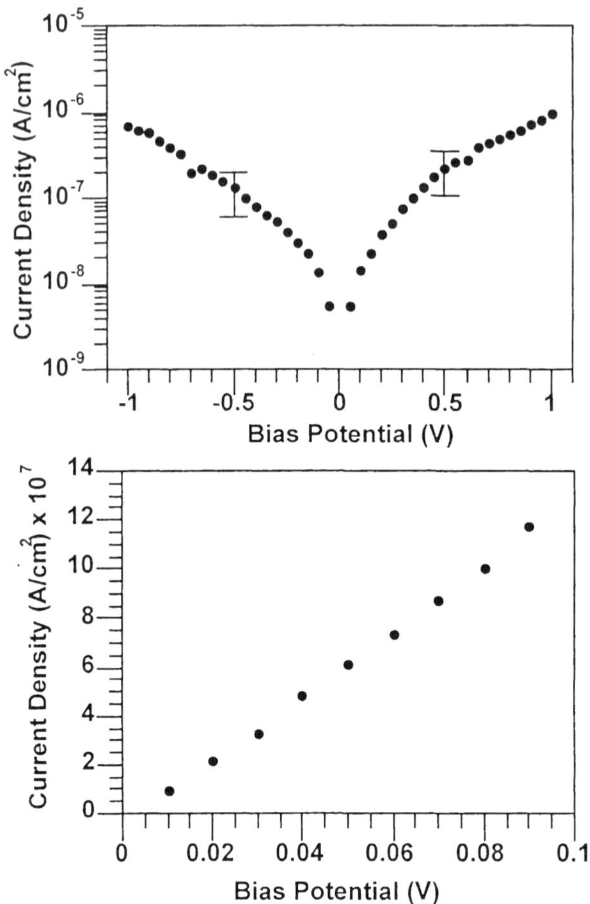

Figure 3. *(A) Plot of current density (logarithmic scale) as a function of bias voltage over the range of –1 to 1 V for a junction with structure $J_{Ag\text{-}C10//C16\text{-}Hg}$. The data are the average of four independent measurements with negative bias and four independent measurements with positive bias. The length of the error bars is representative of the standard deviation obtained from a statistically significant population of junctions (B) Plot of current density (linear scale) as a function of bias voltage over the range of 0-0.1 V for a junction with structure $J_{Ag\text{-}C10//C16\text{-}Hg}$. SOURCE: Reproduced from Reference 58.*

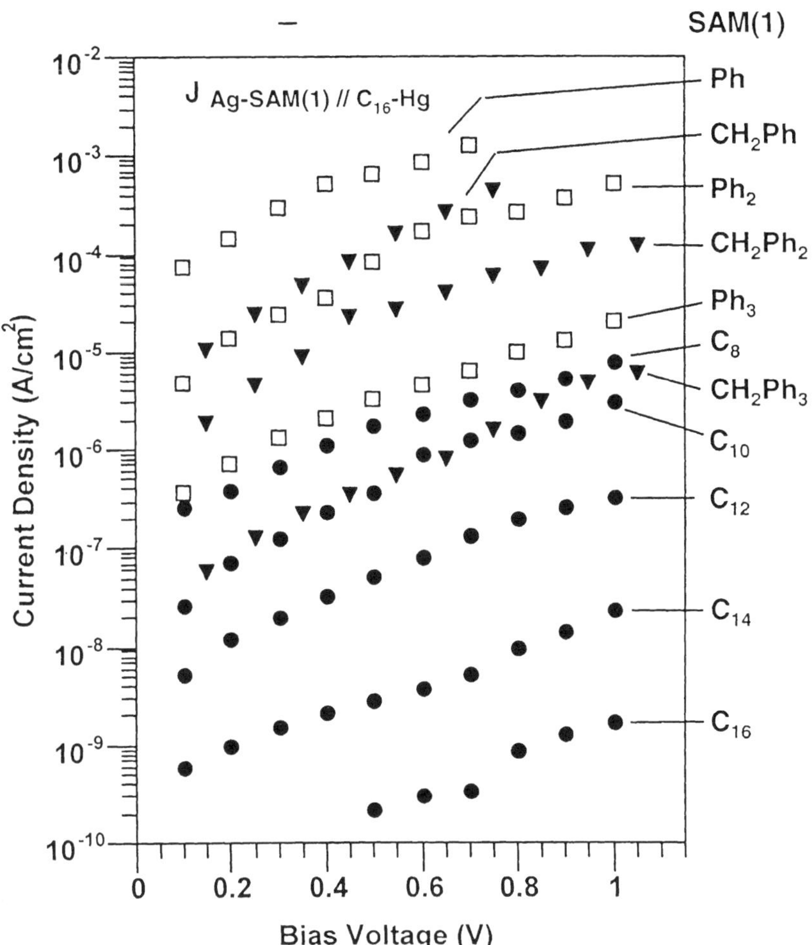

Figure. 4 *Plots of the current density as a function of applied bias potential for the oligomeric SAMs. The aliphatic SAMs were formed from $CH_3(CH_2)_{n-1}SH$ and are represented by C_n. The aromatic SAMs were formed from $C_6H_5(C_6H_4)_{n-1}SH$, reprented by Ph_n, and from $C_6H_5(C_6H_4)_{n-1}CH_2SH$, represented by CH_2Ph_n. SOURCE: Reproduced from Reference 58.*

28

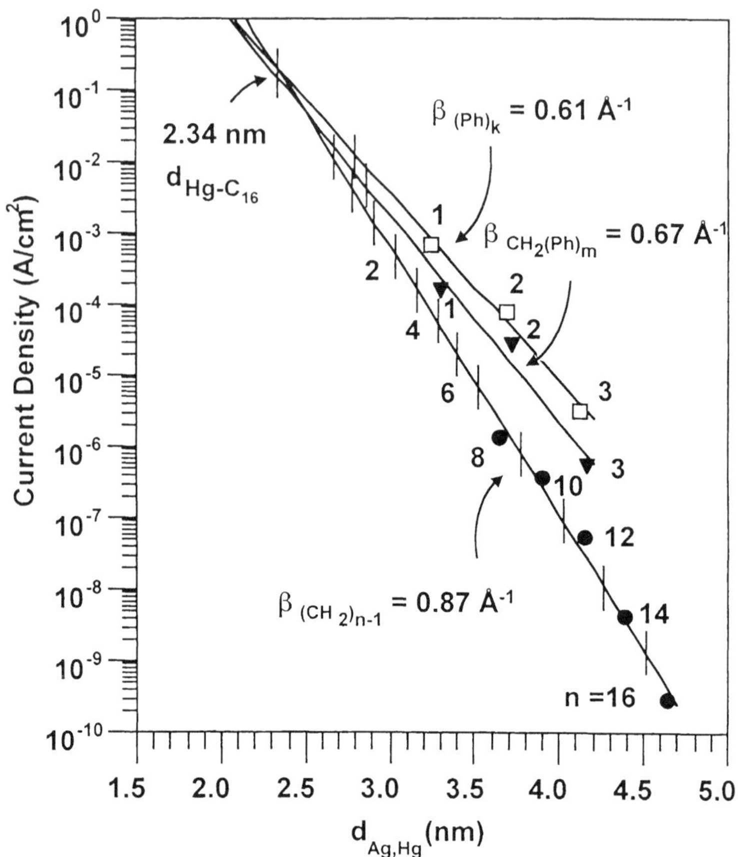

Figure. 5 *Plot of the current density at a potential bias of 0.5 °V as a function of distance for the junctions J $_{Ag\text{-}SAM(1)\,//\,C16\text{-}Hg}$. The aliphatic SAMs were formed from $CH_3(CH_2)_{n\text{-}1}SH$ and are represented by $(CH_2)_{n\text{-}1}$. The aromatic SAMs were formed from $C_6H_5(C_6H_4)_{n\text{-}1}SH$, reprented by Ph_k, and from $C_6H_5(C_6H_4)_{n\text{-}1}CH_2SH$, represented by CH_2Ph_m. SOURCE: Reproduced from Reference 58.*

One of the most interesting features of the I-V data was that distance dependence, β, had little dependence on the voltage applied to the junction (e.g. for junctions formed from aliphatic SAMs, the variation was < 0.1 Å^{-1} over 1 V in bias potential). This result does not agree with the predicted change in β for tunneling through a square barrier.[61] Majda was also unable to fit the voltage dependence of β for alkanethiols SAMs in a Hg-SAM//SAM-Hg junction to this model in agreement with our results.[38] In contrast, Mann and Kuhn successfully fit their I-V data with such a model.[34] Mujica and Ratner have proposed that the electrostatic potential profile in a MIM junctions is nearly flat across the molecular insulator.[62] One of the consequences of this model is that β is nearly independent of bias voltage if electron transport is well-described by superexchange tunneling.[59] Both the shape of our I-V curves and the voltage dependence of β agree with the results of these calculations.

These data demonstrate that the electrical properties measured from this junction are complementary to results obtained from solution-based systems and scanning probe experiments.

Molecular Rectification

One of the challenges in molecular electronics is the fabrication of functional devices. It has been so difficult to fabricate test systems that it has proven impractical to examine a wide range of molecular systems. One device that has drawn particular attention is the molecular rectifier (or a diode). A rectifier is a device that allows a larger flow of current in one bias potential with respect to the opposite bias potential. Rectifiers enable the fabrication of logic circuits and could provide a platform for rudimentary devices for computation.[16]

One strategy for the design of a molecular rectifier is to place an organic molecule comprising a covalently linked electron donor-acceptor pair between two metal electrodes.[63] The hypothesis underlying this strategy is that resonant electron transfer from the acceptor to the donor would enhance current in one bias polarity relative to non-resonant transfer in the opposite bias. Metzger, Sambles, and others have demonstrated experimentally that a system comprising a Langmuir-Blodgett (L-B) film of γ-hexadecyl-quinolinium tricyanoquinomethanide supported on an Al_2O_3/Al electrode and covered by an vapor–deposited Al electrode does indeed rectify current.[37] The synthesis involved in such molecular systems is substantial.

We have studied the current-voltage (I-V) characteristics of junctions consisting of a SAM of tetracyanoquinodimethane (TCNQ) covalently linked to an alkanedisulfide (**1**) on silver or gold and an alkanethiol SAM on mercury.[64] These junctions rectify current although they do not contain a donor-acceptor compound and or an embedded molecular dipole. The rectification ratio, R, (the ratio of the current at 1 V bias in the forward direction to the current in the reverse bias) is ~ 10 if the SAM on mercury is composed of hexadecanthiol (Figure 6). This value of R is close to the value obtained by Metzger (the average value of R based on multiple junctions was 7.55 at 2.2 V and the maximum R for a particular junction was 27.5).[37,41] The current density through the junction can be controlled by changing the length of the alkanethiol on the Hg electrode. The SAM on the Hg electrode also controls the rectification; as the length of the alkyl chain increases, the rectification ratio at 1V decreases.

(1)

The origin of the rectification is still under investigation. A model that is consistent with the data uses the same tight-binding approach that was successful for describing transport in the alkanethiol junctions.[65] In this case, the molecular units bridging the two electrodes are broken into sites consisting of CH_2 groups and the TCNQ moiety. The TCNQ moiety has a different site energy (because of its HOMO-LUMO gap) than the CH_2 groups and also has a different electronic coupling to each electrode (it is covalently linked to the Ag electrode and in van der Waals contact with the SAM on the Hg electrode). The asymmetry in coupling alone can cause a voltage drop across the molecule and a corresponding asymmetry in the current-voltage curve. At this point we have not established which parameter is most important for the TCNQ-based rectifying junction. These results and predictions, nonetheless, suggest a new strategy for the development of molecular rectifiers.

Conclusions

The strengths of metal/SAM//SAM/mercury junctions include the following: (i) They are easy to assemble and use. (ii) They are capable of being used to study a wide range of organic structures, (iii) The range of current that can be measured in them is large (eight orders of magnitude is easily achievable). (iv) There are no ambiguities in the chemical structure and composition of the interfaces resulting from deposition of reactive metals on organic films. (v) They are highly modular because the SAMs on each electrode can be changed

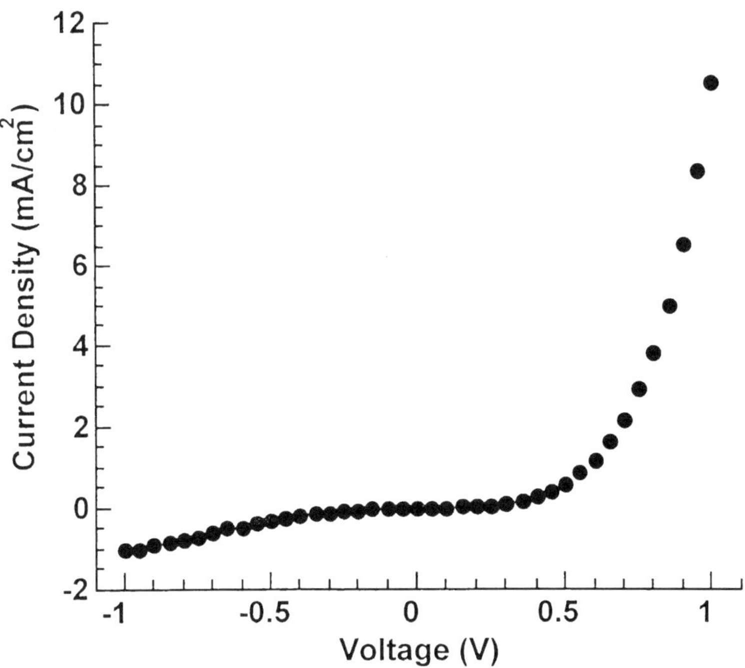

Figure. 6 *Plot of the current density as a function of applied bias potential for a junction with the structure J* $_{Ag\text{-}TCNQ\,/\!/\,C16\text{-}Hg^\cdot}$

independently. (vi) The properties of the two electrodes can be varied widely; by choice of metal (for the planar electrode) and by choice of alloys of liquid metals (for the conformal metal electrode) (vii) Because the fabrication of the junction is simple, it is possible to measure statistically significant numbers of junctions and characterize variations in their properties. (viii) It is possible to disassemble the junction after use and to examine the surface of the planar electrode. The disadvantages of these junctions are: (i) They probably cannot be used to fabricate practical molecular electronic devices. (ii) They do not allow for single molecule resolution achievable in SPM measurement. (iii) Measurement of temperature dependencies is difficult due to the presence of solvent and due to possible changes in the junction near the freezing point of mercury. (iv) There are ambiguities in the influence of the pressure resulting from the applied electrical field on the structure of the SAMs in the junction; for example, the effect of the possible lateral mobility of the SAM on mercury is unknown.[38] The balance of these characteristics is such that we believe that this system represents a valuable new method to study electron transport in MIM junctions. These systems are, thus, very well suited for surveying the properties of ordered, organic monolayer films, and for the physical-organic study of correlations between molecular structure and macroscopic electrical characteristics, and less well suited for high resolution measurements of electron transport.

The work that we have performed so far using the metal/SAM//SAM/mercury junction has demonstrated its utility as a test-bed for molecular electronics. We have measured the electrical breakdown fields of SAMs in these junctions (0.5 GV/m); these fields set the limits for the operating range of potentials that can be applied to these junctions.[54] We have measured the distance dependence of tunneling through aliphatic and aromatic SAMs.[57] These studies substantiated a new model for off-resonance electron transport in MIM junctions.[59] We have also demonstrated that the achievement of molecular rectification only requires the presence of a non-homogenous molecular bridge rather than a D-A molecule. These initial studies have already provide fruitful information about electron transport in MIM junctions.

Acknowledgements

This work was supported by the ONR, DARPA, and the NSF (ECS-9729405). R.E.H. and M.L.C. thank the National Institutes of Health for postdoctoral fellowships and R.H. thanks the Deutsche Forschungsgemeinschaft and the BASF-fellowship program for financial support.

References

1.   Ito, T.; Okazaki, S. *Nature* **2000**, *406*, 1027-1031.
2.   Bennett, C. H.; DiVincenzo, D. P. *Nature* **2000**, *404*, 247-255.

3.   Devoret, M. H.; Schoeklkopf, R. J. *Nature* **2000**, *406*, 1039-1046.

4.   Reed, M. A. *Proc. IEEE* **1999**, *87*, 652-658.

5.   Joachim, C.; Gimzewski, J. K.; Aviram, A. *Nature* **2000**, *408*, 541-548.

6.   Whitesides, G. M.; Mathias, J. P.; Seto, C. T. *Science* **1991**, *254*, 1312-1319.

7.   Ulman, A. *Chem. Rev.* **1996**, *96*, 1533-1554.

8.   Bishop, A. R.; Nuzzo, R. G. *Curr. Opin. Colloid Interface Sci.* **1996**, *1*, 127-136.

9.   Alivisatos, A. P.; Johnsson, K. P.; Peng, X.; Wilson, T. E.; Loweth, C. J.; Bruchez, M. P., Jr.; Schultz, P. G. *Nature (London)* **1996**, *382*, 609-611.

10.   Murray, C. B.; Kagan, C. R.; Bawendi, M. G. *Annu. Rev. Mater. Sci.* **2000**, *30*, 545-610.

11.   Aizenberg, J.; Braun, P. V.; Wiltzius, P. *Phys. Rev. Lett.* **2000**, *84*, 2997-3000.

12.   Hayward, R. C.; Saville, A.; Aksay, A. *Nature* **2000**, *404*, 56-59.

13.   Martin, B. R.; Dermody, D. J.; Reiss, B. D.; Fang, M.; Lyon, L. A.; Natan, M. J.; Mallouk, T. E. *Adv. Mater.* **1999**, *11*, 1021-1025.

14.   Templeton, A. C.; Wuelfing, W. P.; Murray, R. W. *Acc. Chem. Res.* **2000**, *33*, 27-36.

15.   Chen, J.; Calvet, L. C.; Reed, M. A.; Carr, D. W.; Grubisha, D. S.; Bennett, D. W. *Chem. Phys. Lett.* **1999**, *313*, 741-748.

16.   Ellenbogen, J. C.; Love, J. C. *Proc. IEEE* **2000**, *88*, 386-426.

17.   Aviram, A. *J. Am. Chem. Soc.* **1988**, *110*, 5687-92.

18.   Taube, H. *Science* **1984**, *226*, 1028-36.

19.   Closs, G. L.; Miller, J. R. *Science* **1988**, *240*, 440-7.

20.   Gray, H. B.; Winkler, J. R. *Annu. Rev. Biochem.* **1996**, *65*, 537-561.

21.   Marcus, R. A. *Pure Appl. Chem.* **1997**, *69*, 13-29.

22.   Bard, A. J. *Integrated Chemical Systems: A Chemical Approach to Nanotechnology*, 1994.

23.   Murray, R. W. *Acc. Chem. Res.* **1980**, *13*, 135-41.

24.   Chidsey, C. E. D.; Murray, R. W. *Science* **1986**, *231*, 25-31.

25.   Wasielewski, M. R. *Chem. Rev.* **1992**, *92*, 435-461.

26.   Paddon-Row, M. N. *Acc. Chem. Res.* **1994**, *27*, 18-25.

27.   Weber, K.; Hockett, L.; Creager, S. *J. Phys. Chem. B* **1997**, *101*, 8286-8291.

28.   Smalley, J. F.; Feldberg, S. W.; Chidsey, C. E. D.; Linford, M. R.; Newton, M. D.; Liu, Y.-P. *J. Phys. Chem.* **1995**, *99*, 13141-13149.

34

29. Chidsey, C. E. D. *Science* **1991**, *251*, 919-922.
30. Sachs, S. B.; Dudek, S. P.; Hsung, R. P.; Sita, L. R.; Smalley, J. F.; Newton, M. D.; Feldberg, S. W.; Chidsey, C. E. D. *J. Am. Chem. Soc.* **1997**, *119*, 10563-10564.
31. Sikes, H. D.; Smalley, J. F.; Dudek, S. P.; Cook, A. R.; Newton, M. D.; Chidsey, C. E. D.; Feldberg, S. W. *Science* **2001**, *291*, 1519-1523.
32. Fan, F.-R. F.; Yang, J.; Dirk, S. M.; Price, D. W.; Kosynkin, D.; Tour, J. M.; Bard, A. J. *J. Am. Chem. Soc.* **2001**, *123*, 2454-2455.
33. Creager, S.; Yu, C. J.; Bamdad, C.; O'Connor, S.; MacLean, T.; Lam, E.; Chong, Y.; Olsen, G. T.; Luo, J.; Gozin, M.; Kayyem, J. F. *J. Am. Chem. Soc.* **1999**, *121*, 1059-1064.
34. Mann, B.; Kuhn, H. *J. Appl. Phys.* **1971**, *42*, 4398-4405.
35. Vuillaume, D.; Boulas, C.; Collet, J.; Davidovits, J. V.; Rondelez, F. *Appl. Phys. Lett.* **1996**, *69*, 1646-1648.
36. Wong, E. W.; Collier, C. P.; Behloradsky, M.; Raymo, F. M.; Stoddart, J. F.; Heath, J. R. *J. Am. Chem. Soc.* **2000**, *122*, 5831-5840.
37. Metzger, R. M. *Acc. Chem. Res.* **1999**, *32*, 950-957.
38. Slowinski, K.; Fong, H. K. Y.; Majda, M. *J. Am. Chem. Soc.* **1999**, *121*, 7257-7261.
39. Chen, J.; Wang, W.; Reed, M. A.; Rawlett, A. M.; Price, D. W.; Tour, J. M. *Appl. Phys. Lett.* **2000**, *77*, 1224-1226.
40. Collier, C. P.; Matterstei, G.; Wong, E. W.; Luo, Y.; Beverly, K.; Sampaio, J.; Raymo, F. M.; Stoddart, J. F.; Heath, J. R. *Science* **2000**, *289*, 1172-1175.
41. Vuillaume, D.; Chen, B.; Metzger, R. M. *Langmuir* **1999**, *15*, 4011-4017.
42. McCarty, G. S.; Weiss, P. S. *Chem. Rev.* **1999**, *99*, 1983-1990.
43. Datta, S.; Tian, W.; Hong, S.; Reifenberger, R.; Henderson, J. I.; Kubiak, C. P. *Phys. Rev. Lett.* **1997**, *79*, 2530-2533.
44. Bumm, L. A.; Arnold, J. J.; Cygan, M. T.; Dunbar, T. D.; Burgin, T. P.; Jones, L., II; Allara, D. L.; Tour, J. M.; Weiss, P. S. *Science* **1996**, *271*, 1705-07.
45. Bumm, L. A.; Arnold, J. J.; Dunbar, T. D.; Allara, D. L.; Weiss, P. S. *J. Phys. Chem. B* **1999**, *103*, 8122-8127.
46. Salmeron, M.; Neubauer, G.; Folch, A.; Tomitori, M.; Ogletree, D. F.; Sautet, P. *Langmuir* **1993**, *9*, 3600-3611.
47. Wold, D. J.; Frisbie, C. D. *J. Am. Chem. Soc.* **2000**, *122*, 2970-2971.
48. Joachim, C.; Gimzewski, J. K. *Chem. Phys. Lett.* **1997**, *265*, 353-357.
49. Rueckes, T.; Kim, K.; Joselevich, E.; Tseng, G. Y.; Cheung, C.-L.; Lieber, C. M. *Science* **2000**, *289*, 94-97.
50. An analogous experimental system has been used for the study of the capacitance of thin oxide films and is referred to as a "mercury probe." Such devices are commercially available.

51. Any ammeter and stable voltage source can be used as long as the sensitivity and dynamic range of the combination is sufficient.

52. Slowinski, K.; Majda, M. *J. Electroanal. Chem.* **2000**, *491*, 139-147.

53. Whitehead, S. *Dielectric Breakdown of Solids*; Oxford University Press: Oxford, U.K., 1953.

54. Haag, R.; Rampi, M. A.; Holmlin, R. E.; Whitesides, G. M. *J. Am. Chem. Soc.* **1999**, *121*, 7895-7906.

55. Rampi, M. A.; Schueller, O. J. A.; Whitesides, G. M. *Appl. Phys. Lett.* **1998**, *72*, 1781-1783.

56. Zhou, C.; Deshpande, M. R.; Reed, M. A.; Jones, K., II; Tour, J. M. *Appl. Phys. Lett.* **1997**, *71*, 611-613.

57. Holmlin, R. E.; Ismagilov, R. F.; Haag, R.; Mujica, V.; Ratner, M. A.; Rampi, M.; Whitesides, G. M. *Angew. Chem., Int. Ed.* **2001**, *40*, 2316-2320.

58. Holmlin, R. E.; Haag, R.; Chabinyc, M. L.; Ismagilov, R. F.; Cohen, A. E.; Terfort, A.; Rampi, M.; Whitesides, G. M. *J. Am. Chem. Soc.* **2001**, *123*, 5075-5085.

59. Mujica, V.; Ratner, M. A. *Chem. Phys.* **2001**, *264*, 365-370.

60. Fujihara, M.; Inokuchi, H. *Chem. Phys. Lett.* **1972**, *17*, 554-557.

61. Simmons, J. G. *J. Appl. Phys.* **1963**, *34*, 1793-1803.

62. Mujica, V.; Roitberg, A. E.; Ratner, M. *J. Chem. Phys.* **2000**, *112*, 6834-6839.

63. Aviram, A.; Ratner, M. A. *Chem. Phys. Lett.* **1974**, *29*, 277-83.

64. Chabinyc, M. L.; Chen, X.; Holmlin, R. E.; Jacobs, H. O.; Skulason, H.; Frisbie, C. D.; Rampi, M. A.; Whitesides, G. M. Unpublished results.

65. Mujica, V.; Gonzales, C.; Ratner, M. A. Personal communication. .

Chapter 4

Tuning the Electrical Properties of Monolayers by Using Internal Peptide Bonds

Scott M. Reed, Robert S. Clegg, and James E. Hutchison[*]

Department of Chemistry and Materials Science Institute, University of Oregon, Eugene, OR 97403–1253

An approach to the design of molecular electronic materials is presented which is based on adopting design principles from electron transfer proteins. The design of a model system for studying the role of protein structures in mediating electron transfer is discussed. Structural characterization of a self-assembled monolayer containing internal peptide bonds is presented. It is demonstrated that well-ordered monolayers can be formed and measurements of electron transfer rates through these films are described.

Although the rapid development of the electronic technology of this generation has been largely an engineering effort, chemistry plays an important role in this technology. The next generation of electronic devices will be unique, in that chemists will have the opportunity to take a more active role in their development. The scale on which these devices will be built is within the domain of chemistry and it is a challenge to the current generation of chemists to harness the strengths and diversity of chemistry in approaching the problem of developing molecular scale electronic devices.

Electron transfer proteins such as the one shown in Figure 1 provide us with examples of how nanometer sized molecules can control the flow of electrons with great precision. Proteins involved in processes such as photosynthesis and respiration are capable of achieving charge separations over large distances and can direct the flow of electrons within and between proteins with unmatched efficiency and precision (*1*). By modifying proteins such as azurin with metal centers it has been possible to measure electron transfer rates through such proteins (*1c*).

In hopes of learning lessons from these electron-transfer proteins, our work has two main goals. The primary goal is to develop a model system that allows one to isolate and study the structural factors that determine how proteins mediate electron transfer. A second goal is to apply this understanding of the function of proteins to technologically useful platforms.

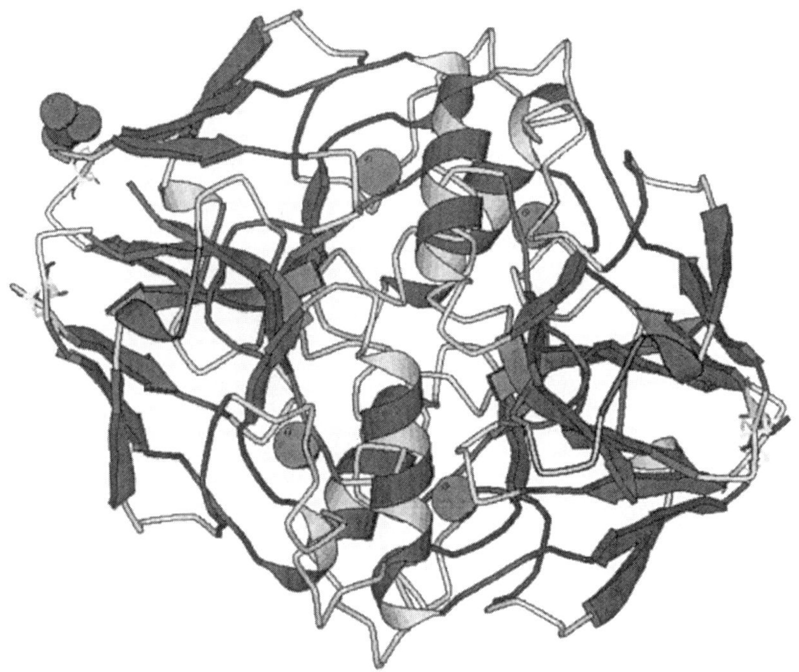

Figure 1. An electron transfer protein, Azurin. Image generated using Molscript (2)

The dependence of long-range electron transfer on the medium separating two redox sites is an important and unresolved problem in biology (*1*). Intramolecular electron transfer between two redox centers has been investigated in modified proteins (*1b,c*) and through polypeptide-based bridging ligands (*3*) in an effort to develop a better understanding of the effects of the intervening protein structure upon biological long-range electron transfer.

Systematic variation of the molecular composition between two redox sites while maintaining a stable, well-defined structure and avoiding intermolecular electron transfer has proven to be difficult however (*3*).

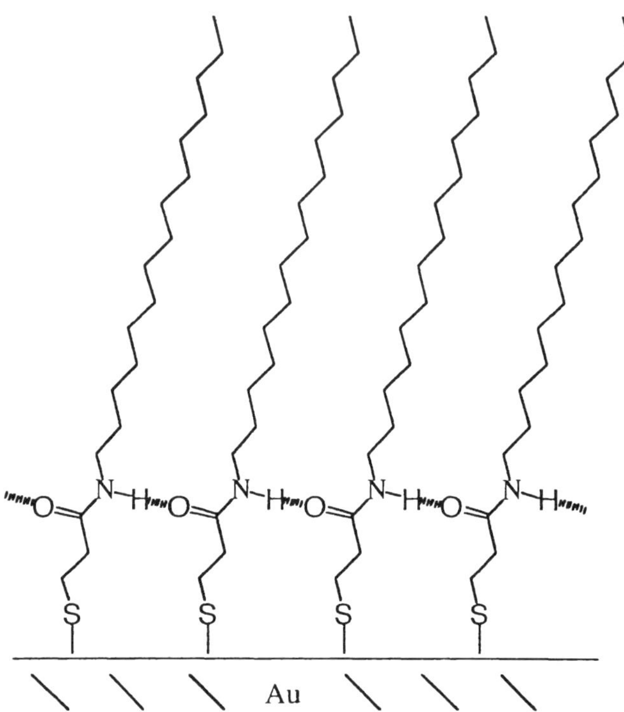

*Figure 2. A peptide-containing alkanethiol self-assembled monolayer (SAM) assembled from 3-mercapto-*N*-dodecylpropionamide (C$_{15}$-1AT/Au)*

We are have designed peptide-containing, alkanethiol self-assembled monolayers (SAMs) on gold (Figure 2), as versatile systems for developing a detailed understanding of how the amino acid composition and structure between redox sites influences electron transfer rates. Alkanethiol monolayers on gold surfaces (*4,5*) offer access to highly ordered, surface-confined molecular structures with a variety of demonstrated applications (*5b,6*), including use as stable, well-defined spacers for electron transfer studies (*5e,7-9*). The densely packed monolayer maintains a precise separation between a gold electrode surface and the pendant redox center and effectively eliminates conformational mobility that can complicate electron transfer rate studies. Through control of the surface concentration of the redox center in the monolayer, intermolecular electron transfer can be eliminated (*8*).

The design of an effective SAM model for the study of protein-mediated electron transfer has several requirements. A synthetic route to the formation of alkanethiols that contain peptide bonds is an essential first step. It is also necessary to design a synthesis that allows terminal functionalization with a redox probe. A third design criterion is that the monolayers must be stable and well ordered to obtain meaningful information from them.

Here we present evidence that we have successfully designed a SAM model useful for the study of electron transfer through peptide-bonds. Evidence from X-ray photoelectron spectroscopy (XPS), external reflectance FTIR spectroscopy (FTIR-ERS), contact angle goniometry (CAG), and cyclic voltammetry (CV) is presented that demonstrates we have designed a system that meets the above criteria. We also describe a general synthetic route to amide-containing alkanethiols that are functionalized with a redox active group. Finally, we will present measurements of the electron transfer rates though peptide-containing SAMs. We will use the following abbreviation C_x-1AT/Au in describing the monolayers, where the 1AT refers to a single amide bond present in the precursor molecule and the x refers to the number of carbon atoms in the alkyl chain above the amide bond.

Experimental

Dichloromethane was distilled over calcium hydride prior to use. Tetrahydofuran was distilled over potassium prior to use. 3,4 dihydro-2*H*-pyran was distilled over sodium. All other reagents were from Aldrich and used without further purification. Thiol deprotections were performed with rigorous exclusion of oxygen. ^{1}H NMR spectra were recorded on a Varion Inova 300 MHz Fourier transform spectrometer. The synthesis of two of the thiols used in this study, 3-mercapto-*N*-dodecylpropionamide (C_{12}-1AT) (*9b*) and 3-mercapto-*N*-pentadecylpropionamide (C_{15}-1AT) have been previously reported (*9a*). The synthesis of the ferrocene terminal Fc-C_{12}-1AT is described next.

Synthesis of cyanoundecanol (**1**). To a solution 11-bromo-1-undecanol (3.581g) in dry DMSO, potassium cyanide (3.672g) is added. The reaction is heated to a reflux for 30 min and then allowed to cool. 50 mL of ice water is added and the product collected by filtration. The product is collected by vacuum distillation as a white solid, 2.61g (92%); mp 37-39° C. ^{1}H NMR (300 MHz, CDCl₃): δ 3.63 (t, 2H), 2.33 (t, 2H), 1.65 (m, 2H), 1.55 (m, 2H), 1.44 (m, 2H), 1.29 (broad m, 12H).

Synthesis of THP protected cyanoundecanol (**2**). To an ice-cold solution of **1** (5.535g) and freshly distilled 3,4 dihydro-2*H*-pyran (12.145g) in dichloromethane (30 mL), p-toluenesulfonic acid (0.584g) is added and the reaction is stirred at 0° for 2 h. The mixture is washed with saturated sodium bicarbonate three times and solvent removed in vacuo, yielding a clear oil, 4.62g (59%); ^{1}H NMR (300 MHz, CDCl₃): δ 4.58 (t, 1H), 3.88 (m, 1H), 3.73 (t, 1H),

3.51 (m, 1H), 3.39 (m, 1H), 2.34 (t, 2H), 1.83 (m, 2H), 1.71 (broad m, 10H), 1.44 (t, 2H), 1.29 (broad m, 12H).

Synthesis of THP protected aminododecanol (3). To a nitrogen sparged mixture of ethyl acetate (60 mL), water (5 mL), and acetic acid (4 mL), platinum oxide (0.04g) is added. Hydrogen gas is bubbled through the solution for 15 min prior to the addition of 2 (0.691g). The reaction is stirred under hydrogen for 12 h and then filtered through celite. Solvent is removed in vacuo, the mixture is redissolved in dichloromethane, washed twice with saturated sodium bicarbonate (100 mL), and the solvent removed in vacuo resulting in a white oily product, 0.700g (100%); ^{1}H NMR (300 MHz, CDCl$_3$): δ 4.59 (t, 1H), 3.88 (m, 1H), 3.73 (t, 1H), 3.53 (m, 1H), 3.39 (m, 1H), 2.89 (t, 2H), 1.83 (m, 2H), 1.59 (m, 6H), 1.26 (broad m, 18H).

Synthesis of THP protected C$_{12}$-1AT-trityl (4). In a 250 mL round bottom flask fitted with a magnetic stir bar, 3 (1.558g) is added to dichloromethane (350 mL) at room temperature. After it has dissolved, 3-N'-dimethylaminopropyl-N-ethylcarbodiimide (1.060g) and 4-dimethylaminopyridine (0.059g) are added. Finally 3-tritylsulfanyl-propionic acid (1.911g) is added. After stirring for 12 h, the organic solution is washed with 3 × 100 mL saturated sodium bicarbonate, 1 × 100 mL distilled water, and 3 × 100 mL 0.01 M hydrochloric acid. The solution is run through a 1 inch plug of basic alumina and the solvent is removed in vacuo. Silica chromatography (hexanes / ethyl acetate) yields a white crystalline product, 0.300g (9%) ^{1}H NMR (300 MHz, CDCl$_3$): δ 7.17-7.45 (m, 15H), 5.66 (t, 1H), 4.58 (t, 1H), 3.87 (m, 1H), 23.74 (m, 1H), 3.50 (m, 1H), 3.38 (m, 1H), 3.14 (m, 2H), 2.49 (t, 2H), 2.03 (t, 2H), 1.58 (broad m, 10H), 1.26 (broad m, 16H).

Synthesis of ω-hydroxy-C$_{12}$-1AT-trityl (5). To a mixture of acetic acid (40 mL), tetrahydofuran (20 mL), and water (10 mL) 4 (0.20g) is added. The reaction is heated to a reflux for 2 h and then allowed to cool. An additional 10 mL of acetic acid is added and the product refluxed for 3 h. After removing solvent and unreacted 4 in vacuo, a white crystalline product is obtained, 0.100g (58%); ^{1}H NMR (300 MHz, CDCl$_3$): δ 7.17-7.43 (m, 15H), 5.30 (broad, 1H), 3.62 (t, 2H), 3.14 (m, 2H), 2.48 (t, 2H), 1.99 (t, 2H), 1.57 (m, 4H), 1.41 (t, 2H), 1.23 (broad m, 14 H).

Synthesis of ω-Fc-C$_{12}$-1AT-trityl (6). Oxalyl chloride (1.0g) is added a 100 mL flask containing ferrocene carboxylic acid (0.080g) under an argon flow. After 2 h stirring at room temperature, the solvent is removed in vacuo. The resultant material is dissolved in freshly distilled dichloromethane and added to a solution of 5 (0.100g). The mixture is stirred for 3 h at room temperature. Solvent is removed in vacuo and the mixture purified by silica chromatography (hexanes / ethyl acetate) yielding a yellow solid, 0.03g (18%); ^{1}H NMR (300 MHz, CDCl$_3$): δ 7.17-7.43 (m, 15H), 5.29 (broad, 1H), 4.79 (t, 2H), 4.37 (t, 2H), 4.18 (s, 5H), 4.04 (t, 2H), 3.14 (m, 2H), 2.48 (t, 2H), 1.99 (t, 2H), 1.58 (m, 4H), 1.40 (t, 2H), 1.24 (broad m, 14H).

Synthesis of Fc-C$_{12}$-1AT (7). Trifluoroacetic acid (3mL) is added to added to a 100 mL flask containing 6 (0.155 g) under nitrogen resulting in an orange solution. Triethylsilane (0.1mL) is added and the bright orange color turns yellow. Solvent is removed in vacuo. Silica chromatography (ethyl acetate / hexanes) yields a yellow crystalline solid, 0.085g (88%); ^{1}H NMR (300 MHz, CDCl$_3$): δ 5.95 (t, 1H), 4.81 (t, 2H), 4.39 (t, 2H), 4.21 (s, 5H), 4.06 (t, 2H), 3.27 (m, 2H), 2.82 (m, 2H), 2.57 (t, 2H), 1.73 (m, 2H), 1.62 (t, 1H), 1.51 (m, 4H), 1.29 (broad m, 14H).

Gold substrates on clean glass microscope slides are prepared by evaporation of 75Å of a chromium adhesion layer followed by 2000Å of gold. The gold substrates are cleaned by ozonolysis for 15 min in a UV-Clean™ (Boekel Industries), rinsed with copious amounts of Nanopure water and degassed absolute ethanol, and dried in a stream of argon immediately prior to formation of monolayers. Contact angles (Nanopure water) were measured on a contact angle goniometer constructed in our laboratory.

SAMs of C$_{15}$-1AT/Au and C$_{12}$-1AT/Au were formed by immersing the gold substrates in 1 mM thiol in degassed absolute ethanol for at least 24 h at room temperature. Electroactive SAMs of C$_{12}$-1AT/Fc-C$_{12}$-1AT/Au are formed by immersing the gold substrates in a solution of 0.1 mM Fc-C$_{12}$-1AT and 0.9 mM C$_{12}$-1AT in degassed absolute ethanol for 24 h at room temperature, followed by ethanol rinsing and soaking in a 0.1 mM solution of C$_{12}$-1AT for 4 days.

FTIR-ERS is performed with polarized light at an 80° angle of incidence using a Spectra-Tech reflectance accessory in a Nicolet Magna-IR 550 Spectrometer, using 1024 signal-averaged scans with Happ-Genzel apodization, 2 cm^{-1} data spacing, and no zero filling. Background spectra are taken on gold substrates coated with 4-mercaptobiphenyl SAMs.

Electrochemical measurements are performed using a BAS 100 B/W electrochemical analyzer. The cell used consists of a gold, SAM-derivatized working electrode, a platinum wire auxillary electrode, and an SSCE reference electrode. For double layer capacitance measurements, the electrolyte is 1.0 M potassium chloride in Nanopure water. For electrochemical blocking studies, the analyte is 1.0 mM potassium ferricyanide. In measurements of the electroactive monolayers, 1.0 M perchloric acid was used as an electrolyte. The geometrical area of the gold working electrode is 0.95 ± 0.03 cm^2.

Results and Discussion

First we will describe a novel synthetic route to electroactive peptide-containing alkanethiols. Next we will describe the structural characterization of C$_{15}$-1AT/Au monolayers by FTIR-ERS, XPS, CAG, and CV demonstrating that these peptide-containing alkanethiols form robust monolayers suitable for electrochemical measurements (*9a*). Electrochemical measurements that were performed on mixed monolayers of FcC$_{12}$-1AT/C$_{12}$-1AT/Au are then reported.

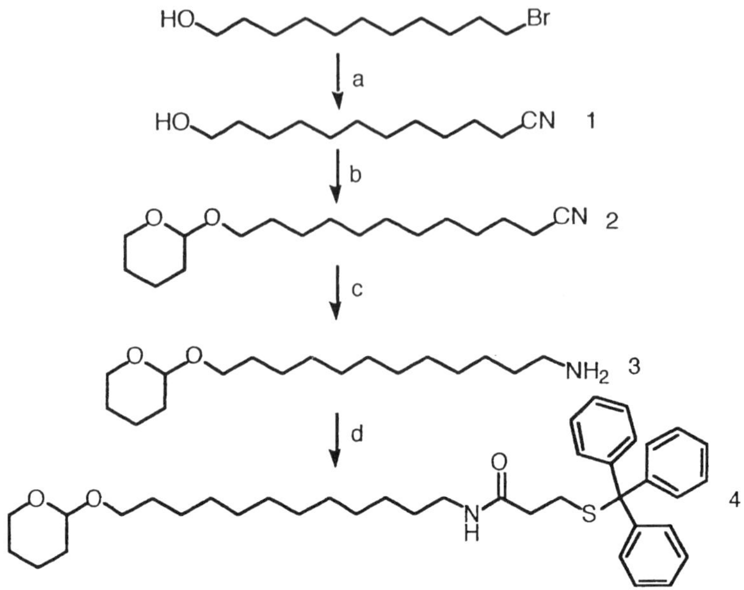

Figure 3. Synthesis of a peptide-containing alkanethiol precursor. a) KCN, DMSO, reflux, 30 min b) DHP, CH₂Cl₂, 0°, 2h c) PtO₂, ethyl acetate, acetic acid d) Tritylsulfanyl-propionic acid, EDCI, DMAP, CH₂Cl₂.

Synthesis

The general synthetic route by which it is possible to synthesize electrochemically addressable peptide-containing alkanethiols is shown in Figures 3 and 4. The synthesis starts with an eleven carbon bromine-terminated alcohol, which is converted to a nitrile **1**. Subsequent protection of the alcohol as a THP ether provides **2** and catalytic reduction of the nitrile results in an amine **3** that can be used in subsequent coupling reactions.

The amine **3** can be coupled to a carboxylic acid, 3-tritylsulfanyl-propionic acid to form the amide bond in **4**. The THP protecting group is easily removed by refluxing in acetic acid to produce the alcohol **5**. No loss of the thiol protecting group occurs during this reaction. Alcohol **5** is coupled to an activated ferrocene carboxylic acid to provide **6**. In the final step, the trityl group is easily removed by trifluoroacetic acid to produce the thiol **7**.

Monolayer Characterization

To determine the suitability of these peptide-containing monolayers for electron transfer experiments, non-electroactive monolayers (Figure 2) were

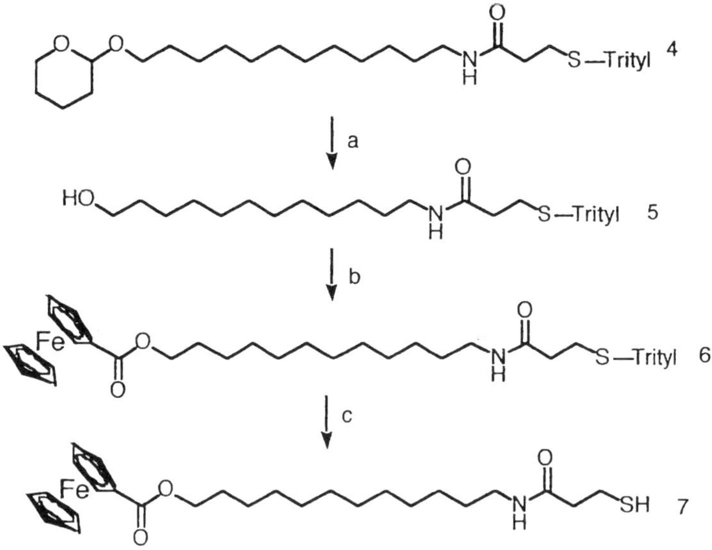

Figure 4. Synthesis of an electroactive peptide-containing alkanethiol a) AcOH, THF, H₂O, reflux, 5h b) Ferrocene carboxylic acid, oxalyl chloride, CH₂Cl₂, rt, 30 min c) TFA, Triethylsilane, rt, 30 min.

characterized by CAG, FTIR-ERS, XPS, and CV (*9a*). The contact angle for water on C_{15}-1AT/Au is $119 \pm 2°$ demonstrating that these SAMs have ordered methyl surfaces comparable to those of well-ordered, octadecanethiol-derived monolayers (C_{18}S/Au), which exhibit contact angles of $118 \pm 2°$ (*5c*).

XPS measurements confirm the presence of oxygen, nitrogen, carbon and sulfur near the gold surface. The binding energies and relative atomic intensities (Table 1) support the monolayer structure shown in Figure 2. The S(2p) peaks are shifted to lower binding energies characteristic of sulfur bound to gold (*5d*). Peaks corresponding to free thiols were not observed (*5d*), indicating that physisorption of C_{15}-1AT or multilayer formation do not occur. Signals from atoms deep in the monolayer (S, N and O) are attenuated (*4b*). As a result, these relative intensities are lower than those calculated from the molecular formula. These XPS data are consistent with the orientation of the monolayer shown in Figure 2. Bare Au shows no detectable C, N, O, or S.

TABLE 1. Composition of C_{15}-1AT/Au by XPS

	BE^a (eV)	atom % by XPS	predicted^b atom %
C(1s$_{alkyl}$), C(1s$_{C=O}$)	285.1, 287.5	90.2	85.7
N(1s)	399.7	4.4	4.8
O(1s)	531.9	3.8	4.8
S(2p$_{3/2}$), S(2p$_{1/2}$)	161.9, 163.1	1.9	4.8

$^a BE$ = binding energy. ^{b}Calculated from the molecular formula.

Evaluation of the crystallinity of the methylene chains was made by using FTIR-ERS (Figure 5) (*10*). The peak frequencies for CH$_2$(as) and CH$_2$(sym) are the same in both the monolayer and in solid C_{15}-1AT and are red-shifted compared to the frequencies for C_{15}-1AT dissolved in CCl$_4$. The red shifts are the same as those seen in the liquid-to-solid phase transition for *n*-alkanes (*11*). The frequencies for the C_{15}-1AT/Au peaks are also the same as those found in crystalline methylene chains such as in C_{18}S/Au monolayers (*4a,5d,10*). The narrow bandwidths for C_{15}-1AT/Au in the high-frequency C–H region (full widths at half maxima: CH$_2$(as), 15 cm^{-1}; CH$_2$(sym), 9 cm^{-1}) are characteristic of tightly packed chains with minimal gauche defects (*12*). Taken together, the peak frequencies and bandwidths are indicative of a microcrystalline array of tightly packed methylene chains in all-*trans* conformations.

The presence of hydrogen bonding in C_{15}-1AT/Au monolayers is evident from FTIR spectra of the amide stretching and bending modes. The nature and extent of hydrogen bonding in the monolayers, as well as the orientation of the amide groups with respect to the substrate can be determined by comparing C_{15}-1AT/Au spectra to data collected for C_{15}-1AT in solution and in the solid phase as well as to appropriate data from the literature.

In a CCl$_4$ solution of C_{15}-1AT (Figure 5, top), where hydrogen bonding is insignificant, the amide A band (N-H stretch) appears as an unassociated peak at 3453 cm^{-1}, amide I (primarily C=O stretch with a small component of C=O-N bend) occurs at 1684 cm^{-1}, and amide II (primarily N-H bend with a small component of C-N stretch) arises at 1511 cm^{-1}. These observations correlate well with published data for secondary amides in solution (*14*).

In solid C_{15}-1AT (Figure 5, middle), amide A shifts to 3302 cm^{-1}, and amide B (first harmonic of N-H in-plane bend) arises as a weak absorption at 3055 cm^{-1}, indicating that hydrogen bonding occurs in solid-phase C_{15}-1AT (*15*). The occurrence of hydrogen bonding in the solid is substantiated by the shift of amide I to 1633 cm^{-1} and amide II to 1541 cm^{-1}, both of which are nearly identical to the amide I (parallel component) and amide II bands seen in Nylon 66. Additionally, these frequencies are similar to those seen in disordered, solid polypeptides (*16*) and are within the range of literature values for solid secondary amides (*14c*). Although solid C_{15}-1AT exhibits hydrogen bonding

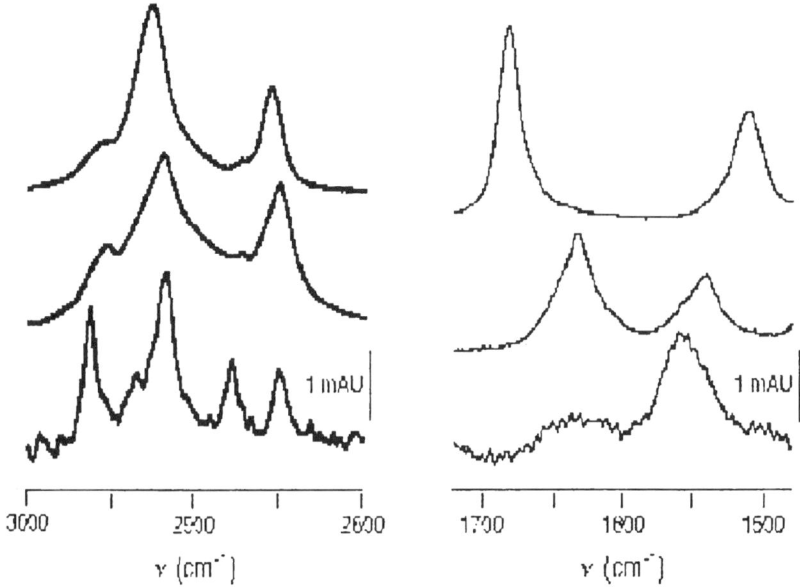

Figure 5. FTIR-ERS spectra of C₁₅-1AT/Au in the high-frequency C–H region (Left). Top: C₁₅-1AT in CCl₄. Middle: solid C₁₅-1AT. Bottom: C₁₅-1AT/Au. Absorbance bar applies to C₁₅-1AT/Au only. FTIR-ERS spectra in the amide I and II region (Right). Top: C₁₅-1AT in CCl₄. Middle: solid C₁₅-1AT. Bottom: C₁₅-1AT/Au. Absorbance bar applies to C₁₅-1AT/Au only. From reference 9a, copyright 1996 American Chemical Society.

between the amide groups, the sample is still disordered as indicated by the breadth of the methylene peak (*12,13*) (Figure 5, middle).

Examination of the amide peaks in the C₁₅-1AT/Au spectrum (Figure 5, bottom) provides compelling evidence of interchain hydrogen bonding and yields information about the average orientation of the amide groups which is consistent with the formation of hydrogen bonds as depicted in Figure 2. The absence of amide A in the reflective spectrum of C₁₅-1AT/Au suggests a parallel orientation of the N-H bond with respect to the surface, and the very broad and unresolved amide I peak likewise infers a similar orientation of the C=O bond (*4b,4d*). In contrast to the aforementioned amide peaks, the amide II appears as a strong peak in the C₁₅-1AT/Au monolayers, confirming the orientation of its dipole with a large component perpendicular to the surface (*4b*). The progression toward higher frequencies from the solution spectrum (1511 cm⁻¹) to the solid (1541 cm⁻¹) to the monolayer (1561 cm⁻¹) suggests an increasing restriction on N-H bending consistent with hydrogen bonding (*16*). As seen in an earlier study of structurally related monolayers (*4b*), the amide II peak exhibits some asymmetry, but for C_{15}-1AT/Au the amide II is much narrower and occurs at a higher frequency. The appearance of a strong amide II peak in concert with the absence of amide A and amide I in the C_{15}-1AT/Au spectra indicates that the dipoles of amide A and amide I are oriented parallel to the gold surface and that the amide is present in the expected *trans* conformation (*4b,4d,16*), as would be required for hydrogen bonding as shown in Figure 2.

Results from FTIR-ERS indicate that C_{15}-1AT/Au possess crystalline, tightly packed methylene chains and extended interchain hydrogen bonding (*9a*). Order has been observed with as few as eleven methylenes in alkanethiol SAMs (*5c*), but templating the hydrocarbon matrix upon the amide moiety increases the threshold value for ordering (*9a*).

Double-layer capacitance measurements allow the films to be probed for their permeability to electrolyte (*5e*). At 100 mV/s scan rate, the charging current for a bare gold electrode is ~15 A at +185 mV. For C_{15}-1AT/Au the charging current at the same potential is reduced to 96 nA. The double-layer capacitance for C_{15}-1AT/Au is 1.0 F/cm^2. This measurement is the same as that found for C_{18}S/Au and indicates low permeability of the monolayer with respect to electrolyte (*18*).

Electrochemical blocking studies give an estimate of possible defect sites, the presence of which could allow facile reduction of ferricyanide (*5e*). At a 100 mV/s scan rate, the peak cathodic Faradaic current at a bare electrode is +205 A. For C_{15}-1AT/Au, the cathodic current at the same potential is reduced to 113 nA, corresponding to a blocking effect of 99.95% (*9*).

In summary, the electrochemical results show that gold electrodes derivatized with C_{15}-1AT exhibit low double-layer capacitance, nearly complete blocking of reduction or oxidation waves in the presence of analyte and an *i-E* response characteristic of electron transfer through a highly ordered barrier containing minimal defects. Taken together with the FTIR, XPS, and contact angle data, these strongly support a highly ordered monolayer with extended hydrogen bonding (*9*).

Gold electrodes derivatized with C_{12}-1AT exhibit low double-layer capacitance, absence of reduction or oxidation waves in the presence of analyte, and an *i-E* response characteristic of electron transfer primarily through an ordered barrier with few defects (*9b*). C_{12}-1AT/Au exhibits a blocking effect of 99.89%. In comparison to C_{15}-1AT/Au, FTIR-ERS of C_{12}-1AT/Au reveals a similar degree of order in the amide region, however, there is evidence of some disorder in the methylene region (*9b*).

Electron Transfer Measurements

An electroactive monolayer is formed by using a mixture of ferrocene-terminal thiols and diluent molecules (amide-containing alkanethiols that do not contain ferrocene) which isolates the ferrocenes from each other. Backsoaking of the monolayer in pure diluent removes ferrocene-containing thiols that are bound at defect sites (*8*). The result is a monolayer that approaches the ideal structure shown in Figure 6.

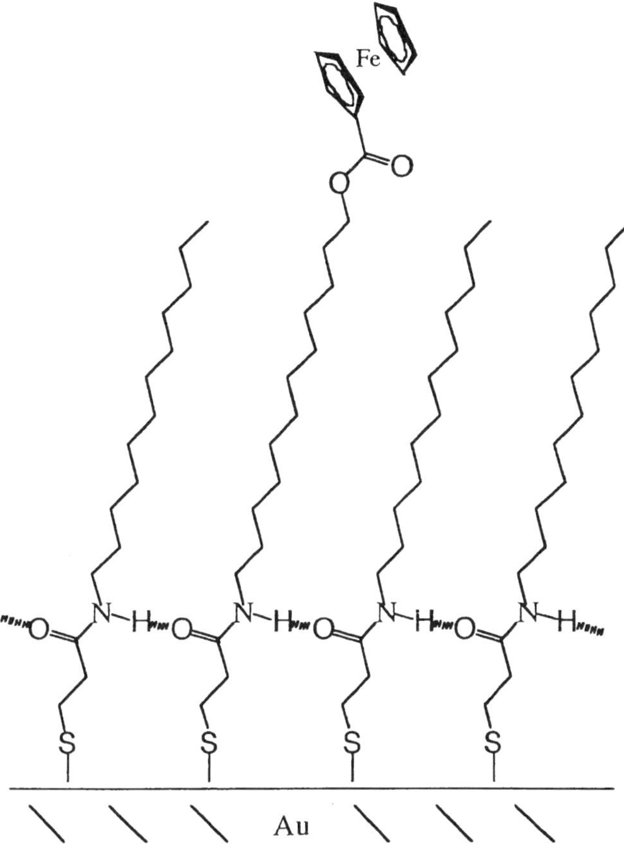

Figure 6. Electroactive peptide-containing C_{12}-1AT/FcC$_{12}$-1AT/Au SAMs

A cyclic voltamogram of C_{12}-1AT/FcC$_{12}$-1AT/Au SAMs is shown in Figure 7. The separation of the anodic and cathodic peaks at a slow scan rate (10 mV/sec) is 20mV. This is indicative of a surface confined species on the electrode surface. The narrowed Ep is consistent with surface-confined electrochemistry. Linear scaling of peak Faradaic current with scan rate is indicative of surface confined electrochemistry. The increased width of the cathodic peak is characteristic of ferrocene carboxylate-terminated monolayers (8), and is caused by differences in oxidized and reduced states of either the monolayer or the solvent sphere surrounding the redox probe. There is no evidence of non-surface-confined electrochemistry. The capacitance of the film where the ferrocene is in the reduced and neutral state provides a good indication of the integrity of the monolayer. The capacitive envelope is narrow and the double layer capacitance of 1.94 F/cm^2 is an indication of a well-ordered film. This capacitance is lower than previously reported (9b) for C_{12}-1AT/Au and may be the result of the longer soak time used to prepare the electroactive films reported here.

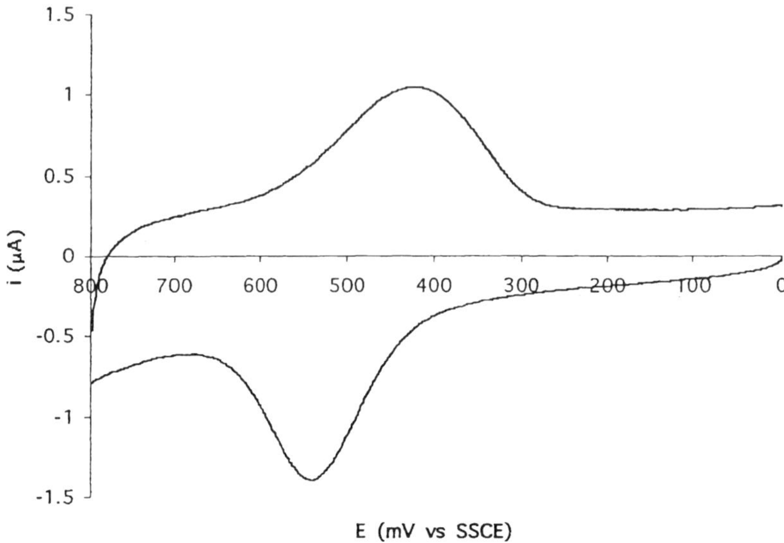

Figure 7. Cyclic voltammogram of C_{12}-1AT/FcC$_{12}$-1AT/Au SAMs at 100 mV/sec in 1 M perchloric acid.

The scan rate dependent peak splitting provides a measure of the electron transfer rate through the peptide-containing monolayer by applying the method of Laviron (*18*). The Laviron method (based on Butler-Volmer kinetics) uses peak splitting that are matched to values on a working curve to obtain a standard electron transfer rate. Re-evaluation of electrode kinetics taking into account Marcus theory (*19,20*) results in substantial changes in the calculated electron transfer rates under conditions where the overpotential exceeds the reorganizational energy. This re-evaluation does not result in substantial changes in the values of electron transfer rate obtained for conditions where the overpotential is smaller than reorganizational energy as in the system described here. This technique is therefore excellent for rapid comparison of similar systems. For this monolayer, the standard electron transfer rate calculated by the Laviron method is 0.7 s^{-1}.

This rate can be compared directly to that measured by Chidsey for an alkanethiol monolayer of the same overall number of atoms in the chain with the same ferrocene carboxylic ester redox probe. In those experiments, the standard electron transfer rate k° obtained by the Laviron method is 1.3 s^{-1} (*7a*) indicating that the inclusion of the amide bond slows the rate of electron transfer through the monolayer. While disorder in the methylene region of the film can affect the electron transfer rate, the capacitive envelope supports the presence of a well-ordered monolayer. The slowing of the electron transfer rate by the inclusion of heteroatoms within the backbone of the monolayer has precedent in the work of Miller (*19*). These observations have been explained by the disruption of the orbital overlap in the coupling pathway through the hydrocarbon chain (*19*).

In conclusion, we have successfully demonstrated that peptide-containing alkanethiol SAMs are robust, well ordered, and function as good electrochemical spacers for the study of electron transfer through peptides. Future work will include examining the effect of various protein structures on electron transfer rates.

Acknowledgement

This research was supported by the National Science Foundation CAREER Program (CHE-9702726), the Camille and Henry Dreyfus Foundation, and a University of Oregon Doctoral Fellowship (SMR). J. E. H. is an Alfred P. Sloan Fellow. We would like to thank Robert Gilbertson for his advice on many aspects of the synthetic route.

Literature Cited

1   (a) Bertini, I.; Gray, H. B.; Lippard, S. J.; Valentine, J. S. *Bioinorganic Chemistry*; University Books: Mill Valley, CA 1994, pp. 315-333. (b) Winkler, J. R.; Gray, H. B. *Chem. Rev.* **1992**, *92*, 369-379. (c) Langen, R.; Chang, I-J.; Germanas, J. P.; Richards, J. H.; Winkler, J. R.; Gray, H. B. *Science* **1995**, *268*, 1733-1735.

2   (a) J. Reichert, A. Jabs, P. Slickers, J. Sühnel *Nucleic Acids Res.* **2000**, 28, 246-249 (b) Per J. Kraulis, *J. App. Crystallography* **1991** *24*, 946-950.

3   Gretchikhine, A. B.; Ogawa, M. Y. *J. Am. Chem. Soc.* **1996**, *118*, 1543-1544.

4   (a) Nuzzo, R. G.; Dubois, L. H.; Allara, D. L. *J. Am. Chem. Soc.* **1990**, *112*, 558-569. (b) Tam-Chang, S.-W.; Biebuyck, H. A.; Whitesides, G. A; Jeon, N.; Nuzzo, R. A. *Langmuir* **1995**, *11*, 4371-4382. (c) Wagner, P.; Hegner, M.; Güntherodt, H.-J.; Semenza, G. *Langmuir* **1995**, *11*, 3867-3875. (d) Lenk, T. J.; Hallmark, V. M.; Hoffman, C. L.; Rabolt, J. F.; Castner, D. G.; Erdelen, C.; Ringsdorf, H. *Langmuir* **1994**, *10*, 4610-4617.

5   (a) Dubois, L. H.; Nuzzo, R. G. *Annu. Rev. Phys. Chem.* **1992**, *43*, 437-463. (b) Nuzzo, R. G.; Allara, D. L. *J. Am. Chem. Soc.* **1983**, *105*, 4481-4483. (c) Bain, C. D.; Troughton, E. B.; Tao, Y.; Evall, J.; Whitesides, G. M.; Nuzzo, R. G. *J. Am. Chem. Soc.* **1989**, *111*, 321-335. (d) Laibinis, P. E.; Whitesides, G. M.; Allara, D. L.; Tao, Y.; Parikh, A. N.; Nuzzo, R. G. *J. Am. Chem. Soc.* **1991**, *113*, 7152-7167. (e) Porter, M. D.; Bright, T. B.; Allara, D. L.; Chidsey, C. E. D. *J. Am. Chem. Soc.* **1987**, *109*, 3559-3568.

6   (a) Kumar, A.; Biebuyck, H. A.; Whitesides, G. M. *Langmuir* **1994**, *10*, 1498-1511. (b) Hutchison, J. E.; Postlethwaite, T. A.; Murray, R. W.

Langmuir **1993**, *9*, 3277-3283. (c) Zak, J.; Yuan, H.; Ho, M.; Porter, M. D. *Langmuir*, **1993**, *9*, 2772-2774.

7 (a) Chidsey, C. E. D. *Science* **1991**, *251*, 919-922. (b) Finklea, H. O.; Hanshew, D. D. *J. Am. Chem. Soc.* **1992**, *9*, 3173-3181.

8 Chidsey, C. E. D.; Bertozzi, C. R.; Putvinski, T. M.; Mujsce, A. M. *J. Am. Chem. Soc.* **1990**, *112*, 4301-4306.

9 (a) Clegg, R. S.; Hutchison J. E. *Langmuir* **1996**, 12, 5239-5243. (b) Clegg, R. S.; Hutchison J. E. *J. Am. Chem. Soc.* **1999**, *121*, 5319-5327. (c) Clegg, R. S.; Reed, S. M.; Hutchison, J. E. *J. Am. Chem. Soc.* **1998**, *120*, 2486-2487.

10 Allara, D. L.; Nuzzo, R. G. *Langmuir* **1985**, *1*, 52-66.

11 Snyder, R. G.; Strauss, H. L.: Elliger, C. A. *J. Phys. Chem.* **1982**, *86*, 5145-5150.

12 Bent, S. F.; Schilling, M. L.; Wilson, W. L.; Katz, H. E.; Harris, A. L. *Chem. Mater.* **1994**, *6*, 122-126.

13 Evans, S. D.; Goppert-Berarducci, K. E.; Urankar, E.; Gerenser, L. J.; Ulman, A. *Langmuir* **1991**, *7*, 2700-2709.

14 (a) Nyquist, R. A. *Spectrochim. Acta A* **1963**, *19*, 509-519. (b) Letaw, H.; Gropp, A. H. *J. Chem. Phys.* **1953**, *21*, 1621-1627. (c) Richards, R. E.; Thompson, H. W. *J. Chem. Soc.* **1947**, 1248-1261.

15 Cha, X.; Ariga, K.; Kunitake, T. *Bull. Chem. Soc. Jpn.* **69**, *16*, 163-168.

16 Miyazawa, T. In *Poly-α-Amino Acids: Protein Models for Conformational Studies*. Fasman, G. D., ed. New York: Marcel Dekker, Inc., 1967.

17 Chen, C.-h.; Hutchison, J. E.; Postlethwaite, T. A.; Richardson, J. N.; Murray, R. W. *Langmuir* **1994** *10*, 3332-3337.

18 Laviron, E. *J. Electroanal. Chem.* **1979**, *101*, 19–28.

19 Tender, L.; Carter, M. T.; Murray, R. W. *Anal. Chem.* **1994** 66, 3173-3781.

20 Weber, K.; Creager, S. E. *Anal. Chem.* **1994** *66*, 3164–3172.

21 (a) Cheng, J.; Sàghi-Szabó, G.; Tossell, J. A.; Miller, C. J. *J. Am. Chem. Soc.* **1996**, *118*, 680–684. (b) Slowinski, K.; Chamberlain, R. V.; Miller, C. J.; Majda, M. *J. Am. Chem. Soc.* **1997**, *119*, 11910–11919.

Chapter 5

Charge–Retention Characteristics of Self-Assembled Monolayers of *Molecular–Wire*-Linked Porphyrins on Gold

Kristian M. Roth[1], Zhiming Liu[1], Daniel T. Gryko[2], Christian Clausen[2], Jonathan S. Lindsey[2], David F. Bocian[2,*], and Werner G. Kuhr[1,*]

[1]Department of Chemistry, University of California, Riverside, CA 92521–0403
[2]Department of Chemistry, North Carolina State University, Raleigh, NC 27695–8204

A molecular approach to information storage employs redox-active molecules attached to an electroactive surface as the memory storage element. Information is stored in the distinct oxidation states of the molecules. Three "molecular-wire" linked porphyrins have been examined each of which forms a self-assembled monolayer (SAM) on a Au microelectrode. The charge-retention time of each "molecular-wire" linked porphyrin mono and dication is considerably shorter than that for phenyl-alkyl chain linked porphyrins.

It is uncertain whether devices that rely on the bulk properties of semiconductor materials will retain the required functionality as feature sizes shrink below 0.1 μm.[1] To overcome the physical limitations of materials used in semiconductor devices, several chemical systems have been investigated with the goal of using discrete molecular properties to replicate the bulk properties of conventional electronic components. Towards this end, molecular switches,[2-4] diodes,[5,6] and wires[7,8] have recently been realized. The integration of molecular components is particularly critical in an information storage system that employs a dynamic random access memory (DRAM).

The basic DRAM cell consists of a capacitor and transistor, where charge is stored to indicate the bit level (0 or 1). This charge decays rapidly in the absence of applied voltage (due to numerous factors), thereby necessitating a periodic refresh. This is the root of the volatile nature of information storage in a DRAM chip. The frequency of the refresh is determined by the magnitude of the capacitor, the voltage at which it was written, and the leakage current. As the feature size of DRAM cells shrink, so does the available capacitance. Recently a DRAM capacitor with a feature size of 0.14 μm has been fabricated with a cell capacitance of ~25 fF.[9] A smaller feature size (< 0.1 μm) will further reduce the effective capacitance to what appears to be a limit based on the bulk properties of doped semiconductors. Only a DRAM based on molecular properties is compatible with a feature size below this limit.

We recently embarked on a program aimed at the design of a molecular-based DRAM.[10] In our approach, a collection of redox-active molecules in a self-assembled monolayer (SAM) attached to an electroactive surface serve as the memory storage element. Information is stored in the distinct oxidation states of the molecules. This approach has the attributes of (1) electrical writing/reading, (2) operation under ambient conditions, (3) low power consumption, (4) no moving parts, (5) reliable operation under multiple cycles, (6) scalability to small dimensions, and (7) fault tolerance through the use of a number of molecules in a given memory storage location.

We have prepared over 100 molecules for studies of molecular-based information storage.[11-17] In each case, the generic architecture consists of the following units: surface attachment group, linker, and redox-active molecules. We have examined a variety of redox-active molecules, including monomeric porphyrins,[11,12] ferrocene–porphyrins,[14] weakly coupled multimeric porphyrins,[15] tightly coupled multimeric porphyrins,[16] and lanthanide triple deckers of porphyrins and phthalocyanines.[17] The emphasis on the multimeric systems and triple deckers was to achieve an increased number of oxidation states, thereby increasing the information storage density. In each case, a thiol group was employed for attachment of the molecules to the surface of Au electrodes. A number of the information-storage molecules were prepared with linkers of different length and composition. The motivation for exploring different linkers was to investigate the effect of linker structure on the duration of charge storage (i.e., charge-retention time) in the SAM after disconnection of the potential applied in the writing process. The ability to retain charge in the absence of applied potential is a key property of this molecular-based approach. A long charge-retention time simultaneously increases the time between refresh cycles (allowing the memory cell to be available for reading and writing a larger percent of the time) and also reduces the total power consumption of the information storage system.

This report focuses on information storage in the multiple oxidation states of porphyrin SAMs on Au surfaces. We have previously shown that SAMs of monomeric porphyrins retain charge for hundreds of seconds after disconnection of the applied potential, almost five orders of magnitude longer than state-of-the-art DRAM elements.[18] Our previous studies of porphyrin monomers bearing phenyl or alkylphenyl linkers revealed that charge-retention times can be altered by ~10-fold via structural modification of the linker. Herein, we conduct similar studies that address the effects of conjugated linkers on charge-retention. The structures of these porphyrins are shown in Figure 1.

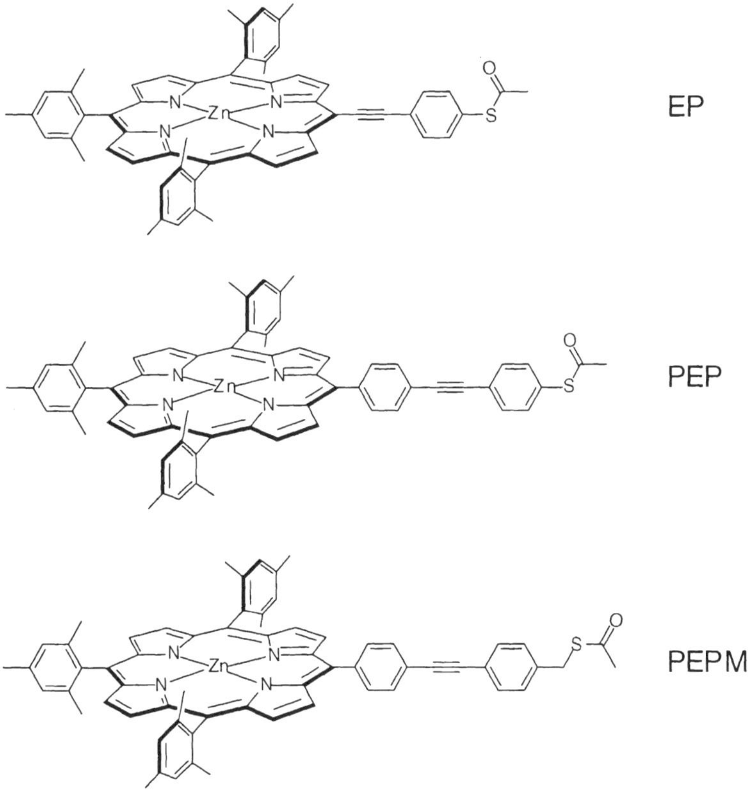

Figure 1. Porphyrins for preparing SAMs with the designations EP (ethynylphenyl linker), PEP (phenylethynylphenyl linker) and PEPM (phenylethynylphenylmethyl linker).

These molecules, designated EP, PEP, and PEPM, differ only with respect to the linker that connects the porphyrin ring and the sulfur atom (ethynylphenyl, phenylethynylphenyl, and phenylethynylphenylmethyl, respectively).

Experimental

The EP, PEP, and PEPM compounds were synthesized as *S*-acetylthio derivatives as previously described.[11] The *S*-acetyl protecting group has been shown to undergo facile cleavage upon exposure to the Au surface.[12,19]

Au ball working electrodes were prepared from 5-μm diameter Au wire (Alpha Aesar) sealed in soft glass. Initially a ~500-μm segment of the wire was protruding from the end of 1-mm i.d. soft glass tubing. When exposed to a flame, the glass forms a tight seal around the Au. The wire exposed to the flame melts into a ball that terminates at the surface of the glass. The electrode was then cooled in a stream of nitrogen and used immediately, resulting in electrodes with an average electrochemical area of 1.5×10^{-4} cm^2.[20]

SAMs were formed by placing the electrode in a 1 mg/mL solution of porphyrin for 5 min. The electrode was then removed from the sample solution and rinsed with distilled CH$_2$Cl$_2$ (Aldrich). All electrochemical potentials, open circuit potential (OCP) measurements, and open circuit potential amperometry (OCPA) measurements were in reference to a bare silver wire (Ag/Ag$^+$) immersed in dried, distilled CH$_2$Cl$_2$ containing 0.1 M Bu$_4$NPF$_6$ (Aldrich, recrystallized three times from methanol and dried under vacuum at 110 °C). All cyclic voltammograms were recorded at 100 V/s with an in-house constructed potentiostat using a routine written in LabVIEW (National Instrument, Austin, TX).

The OCPA measurements were performed as described previously.[10] Initially, the SAM is oxidized with a 20 ms (>> than the RC time constant of the cell) pulse that is ~100 mV above the formal potential of the desired state. The applied potential is disconnected from the counter electrode for a period of time. This disconnection time was varied to evaluate the charge-retention properties. During this time, two events take place: (1) the applied potential is changed to match the empirically determined open circuit potential (OCP) which is ~160 mV vs. Ag/Ag$^+$ and (2) the electrochemical cell relaxes to the OCP. After the disconnect time, the SAM is reduced upon reconnection of the counter electrode and the resulting current is monitored. The magnitude of the observed current is directly proportional to the number of molecules that remain oxidized on the surface while the counter electrode is disconnected. A dramatic reduction in charging current results from the fact that the electrochemical cell is already poised at the OCP before reconnection of the counter electrode. Charge retention is measured by successively changing the disconnect time up to a point where essentially all of the molecules that were initially oxidized have decayed back to the neutral state.

Results and Discussion

Voltammetric Characteristics

The EP, PEP, and PEPM SAMs were characterized by cyclic voltammetry. The electrochemical data for the SAMs are summarized in Table I. The voltametric data indicate that the nature of the linker has very little effect on the $E_{1/2}$ values. A representative cyclic voltammogram of PEPM is shown in Figure 2. The voltammogram of the porphyrin SAM exhibits the rapid and reversible oxidative characteristics that are typical of other porphyrin SAMs we have studied,[10] where each porphyrin SAM exhibits two reversible oxidative waves below 1 V. Repeated voltammetric experiments reveal that these characteristic features do not degrade with successive scans and are stable for hours in solution. The availability of multiple, low-potential oxidation, stable states provides a potential means of storing more than one bit of information in a single storage location, with consumption of relatively little power.

The charge densities of the oxidized SAMs were calculated from the integrated current from the cyclic voltammogram of each molecule, using the electrochemically determined electrode area.[20] The values for PEPM and PEP were comparable, 7.54 and 8.25 $\mu C/cm^2$, respectively, while that for EP was considerably less, 3.22 $\mu C/cm^2$. The lower value for EP is not a result of a larger molecular area, but more likely due to a lower packing density on the surface of the electrode. This is manifested in the voltammogram of EP, which exhibits reduced peak current (which also results in a smaller signal to background charging current ratio). Regardless, the lower surface coverage exhibited by EP does not alter the voltammetric characteristics of this SAM relative to those of PEP or PEPM.

Charge-Retention Characteristics.

Coulometry is the most straightforward technique available for the quantitation of redox species. Faraday's Law {$Q = nFN$, where Q is charge (Coulombs), n is the number of electrons in each reactions, F is Faraday's constant and N is the number of redox molecules (moles)} describes the fundamental relationship between the measured charge (corresponding to the

Table I. Half-wave potentials for thiol-derivatized molecular-wire linked Zn porphyrins[a]

	Half-wave potential (V)		*Charge density ($\mu C/cm^2$)*
Porphyrin	$E_{0 \to 1}$	$E_{+1 \to 2}$	σ^{b}
PEPM	0.58	0.93	7.54
PEP	0.60	0.97	8.25
EP	0.59	0.98	3.31

[a]Obtained in CH_2Cl_2 containing 0.1 M Bu_4NPF_6 vs. Ag/Ag^+; scanrate = 100 V/s.

[b]Charge density ($\mu C/cm^2$) determined from cyclic voltammetry.

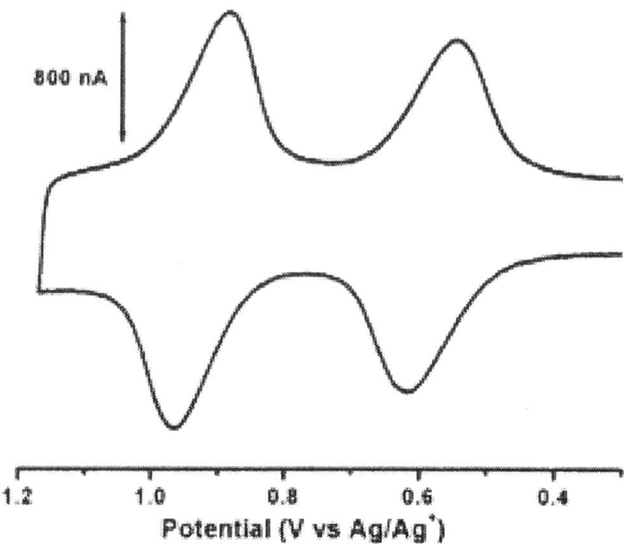

Figure 2. Cyclic staircase voltammetry (100 V s⁻¹) of PEPM SAM on a Au ball electrode immersed in a solution of 0.10 M Bu₄NPF₆ in dried, distilled CH₂Cl₂ using Ag wire as a counter electrode.

oxidation or reduction of the redox species) and the number of molecules consumed in the redox process. The only assumption in this relationship is that all the measured charge results from faradaic processes and not due to charging of the electrochemical double layer. Thus, it is necessary to minimize the background charging current to accurately determine the number of molecules participating in the redox process. One way to accomplish this is to discharge the charging current prior to the measurement of the charge associated with a redox process.

Measurement of stored charge in a SAM has been accomplished using the OCPA technique.[10] This technique requires that (1) the OCP of the electrochemical cell can be readily determined, and is at an analytically useful potential; (2) the molecules in the SAM remain oxidized under open circuit conditions; and (3) the molecules that remain oxidized are readily reduced upon reconnection of the counter electrode at the OCP. As previously shown,[10,13,18] SAMs of various porphyrins satisfy these conditions.

Representative data for the OCPA of PEPM are shown in Figure 3A. The traces are the reductive current required to neutralize the mono-oxidized SAM ($E_{applied}$ = 0.8 V). The different traces were measured at various times following the reconnection of the counter electrode. The current transients are integrated to yield the number of molecules that remain oxidized after the disconnect time (Figure 3B). The integrated charge measured after 80 μs represents essentially the total retained charge of the SAM. For all three molecules, the decay in the total charge as a function of disconnect time fits a first-order rate law with a high fidelity ($r^2 = 0.98$) for both oxidation states ($E_{+2} \rightarrow E_0$ and $E_{+1} \rightarrow E_0$). Simple exponential decay is not expected for the decay of the second state owing to the fact that it presumably decays via a sequential process ($E_{+2} \rightarrow E_{+1} \rightarrow E_0$). Nevertheless, we have also observed pseudo-first order decay for the higher oxidation states of other porphyrin SAMs.[13]

The (pseudo) first-order charge-retention kinetics exhibited by the porphyrin SAMs allows the determination of a half-life ($t_{1/2}$) for charge retention. The $t_{1/2}$ values for the EP, PEP, and PEPM SAMs are listed in Table II. The charge at time zero (σ_0) is extrapolated from the fits of the decay curves and corresponds to the total charge in the SAM at a disconnect time of zero seconds. This value represents the amount of stored charge (oxidized molecules) in the SAM prior to any decay process and correlates very well with the integrated voltammograms (as can be seen in Tables I and II). The values are given as charge densities to facilitate comparison of different electrode sizes. The good agreement between the values indicates that the charge observed in OCPA is well correlated to the faradaic current measured with cyclic voltammetry.

Inspection of the $t_{1/2}$ values shown in Table II reveals that EP exhibits the shortest charge-retention characteristics; the $t_{1/2}$ for the first oxidation state of this SAM is ~31 s. The $t_{1/2}$ values become successively longer for PEP and PEPM and are 43 and 58 s, respectively. The $t_{1/2}$ values for the second oxidation states of the SAMs follow the same trend but are systematically longer than those of the first oxidation state. The longer $t_{1/2}$ for the higher oxidation states are consistent with our studies on other multiply oxidized porphyrin SAMs.[13]

The monotonically increasing $t_{1/2}$ values, EP < PEP < PEPM, are qualitatively consistent with the increasing length of the linker in the three molecules. However, the correlation between the $t_{1/2}$ values and actual linker length is poor. In particular, the $t_{1/2}$ value(s) for PEP are only about 30% longer than those for EP, despite the fact that the additional phenyl ring in the linker of PEP results in a tether length that is approximately double that of EP. An equivalent (or larger) increase in $t_{1/2}$ value is observed for PEPM versus PEP, where a much shorter methylene group has been added to the linker. These observations suggest that factors other than tether length may affect the $t_{1/2}$ values. In this connection, it is interesting that the σ_0 value for EP is significantly less (approximately half) than that for PEP or PEPM (Tables I and II). As noted above, the lower charge density for EP indicates lower surface coverage. The lower surface coverage could reflect either a different tilt angle of the molecules with respect to the surface or different packing arrangement (or both). The distance of the electroactive molecule from the surface and/or the molecular packing could also influence the $t_{1/2}$ value.

Results from an earlier study by our group shed additional light on the effects of tether length on charge retention. In this study, we investigated the effects of methylene spacers on the $t_{1/2}$ values of a series of analogous porphyrin SAMs containing phenyl-$(CH_2)_n$ linkers, where $n = 0 - 3$ (termed PMn), shown in Figure 4. The molecular area occupied by each of these molecules is similar suggesting similar tilt angles and molecular packing arrangements. The $t_{1/2}$

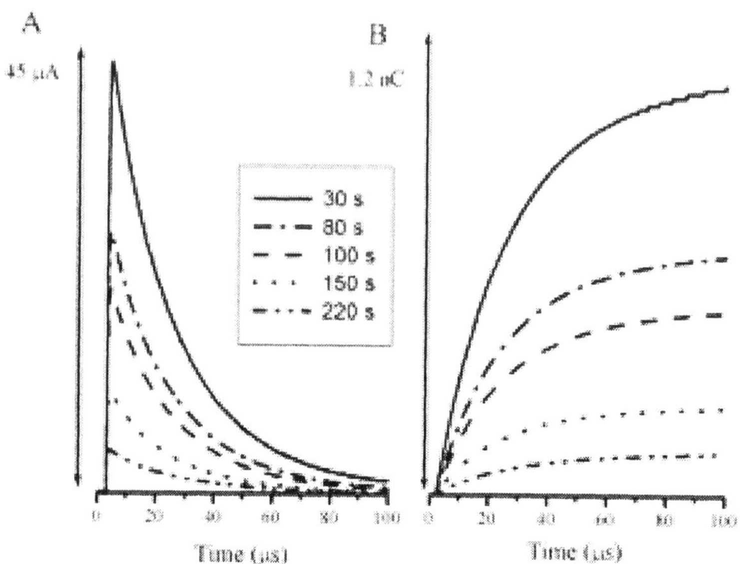

Figure 3. (A) Au electrode with a PEPM SAM was oxidized (poised at 0.8 V vs. Ag/Ag⁺ for 20 ms) then disconnected from the applied potential; (A) each transient results from the reconnection of the counter electrode at the empirically determined OCP (for a PEPM SAM on Au the OCP is –165 mV vs. Ag/Ag⁺) at 5 unique disconnect times ranging from 30 – 220 s; (B) the integration of the transients in (A).

Table II. Charge densities and charge-retention half-lives for thiol-derivatized molecular-wire linked Zn-porphyrins in SAMs[a]

Porphyrin	State 1		State 2	
	$t_{1/2}$	σ_0	$t_{1/2}$	σ_0
PEPM	58	7.40	94	13.9
PEP	43	8.39	66	16.6
EP	31	4.30	59	8.73

[a]Charge density at t = 0 s ($\mu C/cm^2$) and charge-retention half-life ($t_{1/2}$/s) obtained by fitting the decay in the observed charge vs. disconnect time to a first-order rate law.

values for PMn SAMs were found to increase monotonically with the successive addition of each methylene group. In the case of the first oxidation state, these values are 116, 167, 656, and 885 s, respectively.[10] Accordingly, when the molecular packing of the SAMs is similar, the $t_{1/2}$ values appear to be well-correlated with the tether length. Another noteworthy observation is that the $t_{1/2}$ values for the PMn SAMs are longer than those of EP, PEP or PEPM, despite the fact that linkers for the former SAMs are shorter than those of the latter (except perhaps in the case of EP versus PM3).

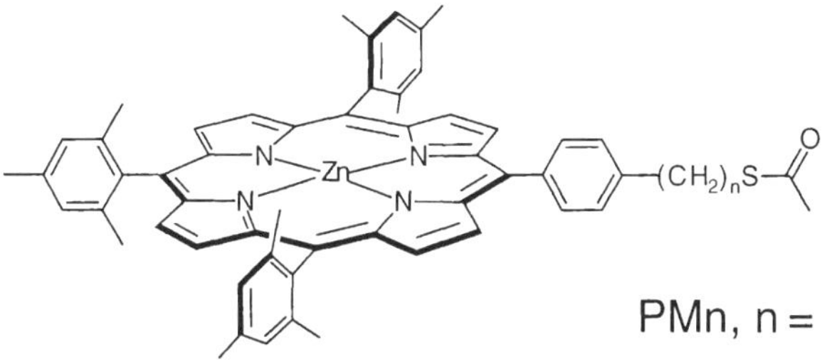

PMn, n = 0-3

Figure 4. Phenylalkyl-linked porphyrins studied previously in SAMs.

The studies reported herein indicate that the nature of the linker plays a significant role in determining the charge-retention characteristics of the porphyrin SAMs. The length of the linker is clearly important, with longer linkers generally giving longer $t_{1/2}$ values. However, the relative $t_{1/2}$ values of the "wire-linked" porphyrins studied here versus those of the PMn SAMs studied previously show that the electronic properties of the linker have a strong influence on charge retention. The fact that the $t_{1/2}$ values for the "wire-linked" porphyrin SAMs are significantly shorter than those of the PMn SAMs is qualitatively consistent with greater electronic coupling between the Au surface and the porphyrin redox site through the more highly conjugated linkers of the wire-linked molecules. However, the detailed nature of the electronic coupling is clearly more complicated. In particular, there is no obvious rationale for why the $t_{1/2}$ value for the PM0 SAM, which contains a single phenyl linker, should be significantly longer than for the "wire-linked" porphyrin SAMs studied herein. The phenyl linker in PM0 should arguably facilitate electronic coupling comparable to (or stronger) than linkers such as diphenylethyne.

Finally, we close by noting that the charge-retention times for all of the porphyrin SAMs that we have studied reflect rates of charge recombination that are much longer than the rates of forward electron transfer that are characteristic of the oxidation process. This assessment is based on the observation that the current-transient response measured during oxidation is limited by the RC time constant of the electrochemical cell, not the intrinsic rate of electron transfer.[10] Accordingly, we do not know whether the rates of forward electron transfer in the porphyrin SAMs exhibit linker-dependent trends that parallel the rates of charge recombination. In this connection, however, the rates of forward electron transfer have been examined for SAMs of ferrocenes tethered by a variety of linkers. The forward electron-transfer rates for alkylferrocene SAMs monotonically decrease as the length of the alkyl chain increases.[21-23] In addition, the rates of electron transfer through conjugated linkers (such as oligophenylethyne) are much faster than through alkyl linkers.[24-26] These trends are generally consistent with those we have reported herein and previously for charge retention (i.e., the rates of charge recombination decrease as methylene spacers are added to the linker and are generally faster for conjugated than for non-conjugated linkers).[10] We also note that in a recent study, the rates of forward electron transfer in a series of oligophenylene-vinylene-tethered ferrocenes were shown to exhibit only a weak dependence on the number of conjugating elements in the linker (and therefore on the length of the linker).[27] This behavior is qualitatively similar to our observation that the rates of charge recombination in the "wire-linked" porphyrin SAMs are not appreciably different for phenylethyne and diphenylethyne linkers. Regardless, a more detailed analysis of the correlation between the rates of forward electron transfer and charge recombination as a function of linker must await further studies wherein the rates of forward electron transfer are explicitly measured for a series of porphyrin SAMs. Gaining a fundamental understanding of the relationship between linker structure and electron-transfer rates is essential for the rational design of molecules for specific information-storage applications.

Acknowledgments

This work was supported by the DARPA Moletronics Program, administered by the ONR (Grant No. N00014-99-1-0357) and Zettacore, Inc.

References

1. Harrell, S.; Seidel, T.; Fay, B. Microelectron. Eng. 1996, 30, 11-15.
2. Collier, C. P.; Matterstei, G.; Wong, E. W.; Luo, Y.; Beverly, K.; Sampaio, J.; Raymo, F. M.; Stoddart, J. F.; Heath, J. R. Science (Washington, D. C.) 2000, 289, 1172-1175.
3. Wong, E. W.; Collier, C. P.; Behloradsky, M.; Raymo, F. M.; Stoddart, J. F.; Heath, J. R. J. Am. Chem. Soc. 2000, 122, 5831-5840.
4. Collier, C. P.; Wong, E. W.; Belohradsky, M.; Raymo, F. M.; Stoddart, J. F.; Kuekes, P. J.; Williams, R. S.; Heath, J. R. Science (Washington, D. C.) 1999, 285, 391-394.
5. Chen, J.; Wang, W.; Reed, M. A.; Rawlett, A. M.; Price, D. W.; Tour, J. M. Appl. Phys. Lett. 2000, 77, 1224-1226.

6. Chen, J.; Wang, W.; Reed, M. A.; Rawlett, A. M.; Price, D. W.; Tour, J. M. Mater. Res. Soc. Symp. Proc. 2001, 582, H3.2/1-H3.2/5.
7. Mallouk, T. E. Science (Washington, DC) 2001, 291, 443-444.
8. Yu, J. S.; Kim, J. Y.; Lee, S.; Mbindyo, J. K. N.; Martin, B. R.; Mallouk, T. E. Chem. Commun. 2000, 2445-2446.
9. Yoon, H.; Cha, G. W.; Yoo, C.; Kim, N. J.; Kim, K. Y.; Lee, C. H.; Lim, K. N.; Lee, K.; Jeon, J. Y.; Jung, T. S.; Jeong, H.; Chung, T. Y.; Kim, J.; Cho, S. I. IEEE J. Solid-State Circuits 1999, 34, 1589-1599.
10. Roth, K. M.; Dontha, N.; Dabke, R. B.; Gryko, D. T.; Clausen, C.; Lindsey, J. S.; Bocian, D. F.; Kuhr, W. G. J. Vac. Sci. Technol. B 2000, 18, 2359-2364.
11. Gryko, D. T.; Clausen, C.; Roth, K. M.; Dontha, N.; Bocian, D. F.; Kuhr, W. G.; Lindsey, J. S. J. Org. Chem. 2000, 65, 7345-7355.
12. Gryko, D. T.; Clausen, C.; Lindsey, J. S. J. Org. Chem. 1999, 64, 8635-8647.
13. Gryko, D.; Li, J.; Diers, J. R.; Roth, K. M.; Bocian, D. F.; Kuhr, W. G.; Lindsey, J. S. J. Mater. Chem. 2001, 11.
14. Gryko, D. T.; Zhao, F.; Yasseri, A. A.; Roth, K. M.; Bocian, D. F.; Kuhr, W. G.; Lindsey, J. S. J. Org. Chem. 2000, 65, 7356-7362.
15. Clausen, C.; Gryko, D. T.; Dabke, R. B.; Dontha, N.; Bocian, D. F.; Kuhr, W. G.; Lindsey, J. S. J. Org. Chem. 2000, 65, 7363-7370.
16. Clausen, C.; Gryko, D. T.; Yasseri, A. A.; Diers, J. R.; Bocian, D. F.; Kuhr, W. G.; Lindsey, J. S. J. Org. Chem. 2000, 65, 7371-7378.
17. Li, J.; Gryko, D.; Dabke, R. B.; Diers, J. R.; Bocian, D. F.; Kuhr, W. G.; Lindsey, J. S. J. Org. Chem. 2000, 65, 7379-7390.
18. Roth, K. M.; Lindsey, J. S.; Bocian, D. F.; Kuhr, W. G. Anal. Chem. submitted.
19. Tour, J. M.; Jones, L., II; Pearson, D. L.; Lamba, J. J. S.; Burgin, T. P.; Whitesides, G. M.; Allara, D. L.; Parikh, A. N.; Atre, S. J. Am. Chem. Soc. 1995, 117, 9529-9534.
20. Bard, A. J.; Faulkner, L. R. Electrochemical Methods: Fundamentals and Applications; John Wiley & Sons: New York, 1982.
21. Creager, S.; Yu, C. J.; Bamdad, C.; O'Connor, S.; MacLean, T.; Lam, E.; Chong, Y.; Olsen, G. T.; Luo, J.; Gozin, M.; Kayyem, J. F. J. Am. Chem. Soc. 1999, 121, 1059-1064.
22. Chidsey, C. E. D. Science (Washington, D. C.) 1991, 251, 919-22.
23. Smalley, J. F.; Feldberg, S. W.; Chidsey, C. E. D.; Linford, M. R.; Newton, M. D.; Liu, Y.-P. J. Phys. Chem. 1995, 99, 13141-13149.
24. Adams, R. D.; Barnard, T.; Rawlett, A.; Tour, J. M. Eur. J. Inorg. Chem. 1998, 429-431.
25. Hsung, R. P.; Chidsey, C. E. D.; Sita, L. R. Organometallics 1995, 14, 4808-4815.
26. Sachs, S. B.; Dudek, S. P.; Hsung, R. P.; Sita, L. R.; Smalley, J. F.; Newton, M. D.; Feldberg, S. W.; Chidsey, C. E. D. J. Am. Chem. Soc. 1997, 119, 10563-10564.
27. Sikes, H. D.; Smalley, J. F.; Dudek, S. P.; Cook, A. R.; Newton, M. D.; Chidsey, C. E. D.; Feldberg, S. W. Science (Washington, D.C.) 2001, 291, 1519-1523.

Chapter 6

Effect of Molecular Properties on Electron Transmission through Organic Monolayer Films

A. M. Napper[1], Haiying Liu[1], H. Yamamoto[1], D. Khoshtariya[1,2],
and D. H. Waldeck[1,*]

[1]Chemistry Department, University of Pittsburgh, Pittsburgh, PA 15260
[2]Permanent address: Institute of Molecular Biology and Biophysics, Georgian
Academy of Sciences, Gotua 14, Tbilisi 380060, Georgian Republic
[*]Corresponding author: email: dave@pitt.edu

ABSTRACT: This work explores how the structure, composition, and geometry of organic thin films impact the tunneling of electrons from an electrode to a redox species in solution. The impact of the film's thickness on the electronic coupling strength and the transition between weak and strong coupling is discussed first. The influence of film composition is explored by comparing the electron transfer rates through alkane and ether linked redox couples. The last section of the manuscript describes the impact of an organic chain's tilt angle on the decay length for the electron transfer rate constant.

Electron transfer through insulating organic monolayer films has been of much current interest [1]. Such systems are very useful for investigating fundamental issues in electron transfer and are expected to be very important in molecular electronics and bioelectronics. To control electron flow on nanometer length scales it is critical to understand and manipulate both the conducting and insulating properties of molecules and molecular assemblies through variation of structural and compositional parameters. Although considerable work has been performed in this area, the frontiers are still being explored. This manuscript describes three recent studies performed by our group on electron transfer through ultrathin organic films. The first part of the manuscript describes the general physical view of the electron transfer process and the dependence of the electron transfer rate constant on the system's control parameters. Our recent studies that explore the transition between the adiabatic and nonadiabatic electron transfer regimes with the organic film's thickness and the solution viscosity are discussed here. The second part of the manuscript describes a recent study that explores how the electron transfer depends on the film composition. In particular, we demonstrate that the replacement of a methylene link in an alkane chain by an ether linkage reduces the electron transfer rate constant by almost five times because of the change in the electronic nature of the tunneling barrier. This work may have important implications for future device design. The third part of the manuscript explores the nature of the electron tunneling through films and whether it depends on superexchange through covalent linkages or involves intermolecular couplings. By compiling data from a range of sources it is possible to examine how the electron transfer's distance dependence changes with the angle of tilt between the insulating chains and the electrode surface.

Mechanism of Electron Transfer: Figure 1 shows a schematic diagram for the physical picture of electron transfer through insulating films on electrode surfaces. The top part of the figure illustrates the structure of the interface and consists of the metal or semiconductor electrode on the left, the organic monolayer film in the middle, and a redox couple on the right. The block arrow is

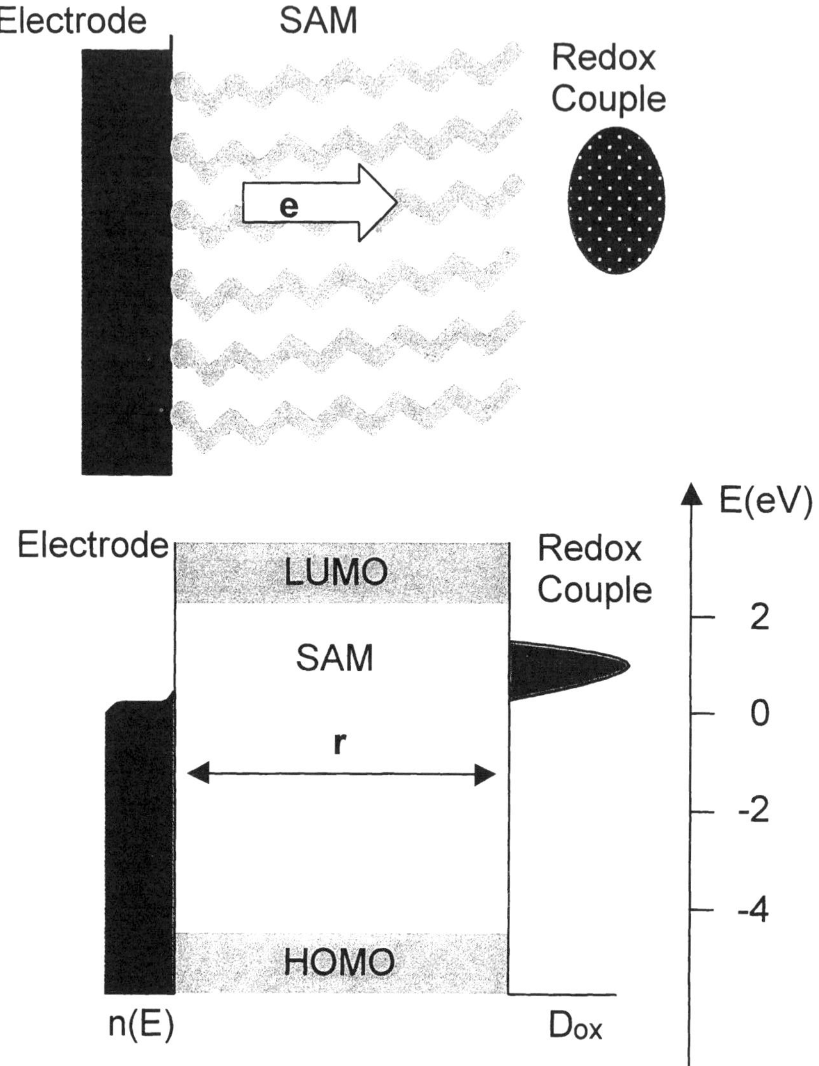

Figure 1: The top diagram illustrates the physical structure of the devices, containing an electrode, an alkanethiol monolayer, and a redox couple. The bottom diagram provides the electronic energy scale for these different regions.

meant to represent an electron moving through the interface. The bottom diagram sketches the electronic energy scales associated with the different regions of the structure. The electrodes have typical work functions of 4 to 5 eV and the ionization potential of the insulating films (alkane and ether) is typically > 8 or 9 eV. The electron transfer rate constant depends on the availability of electronic states (at the electrode and in the redox couple) and the probability of tunneling between them.

When the thickness of the insulating barrier is large, the electronic interaction between the redox couple and the electrode is weak and the electron transfer occurs in the nonadiabatic limit [2]. In this case the rate constant k^{NA} may be written as

$$k_{red}^{NA}(\eta) = \int \rho(\varepsilon) \cdot f(\varepsilon) \cdot D_{ox}(\varepsilon, \lambda, \eta) \cdot P(\varepsilon, \eta) \cdot d\varepsilon, \qquad 1$$

in which the integral is taken over the electronic energy ε. The available electronic states in the electrode are determined by their density distribution $\rho(\varepsilon)$ and the Fermi-Dirac distribution law $f(\varepsilon)$. The term $D_{ox}(\varepsilon,\lambda,\eta)$ represents the density of electron accepting levels (for an oxidation reaction this term would be the density of electron donating levels), and $P(\varepsilon,\eta)$ is the tunneling probability. The parameter λ is the reorganization energy for the redox couple and η is the overpotential. Using the classical formulated Marcus model, this rate expression can be cast as

$$k_{red}^{NA}(\eta) = \frac{2\pi}{\hbar} |V|^2 \frac{1}{\sqrt{4\pi\lambda k_B T}} \int_{-\infty}^{\infty} \rho(\varepsilon) \cdot f(\varepsilon) \cdot \exp\left[-\left(\frac{(\lambda + (\varepsilon_F - \varepsilon) + e\eta)^2}{4\lambda k_B T}\right)\right] \cdot d\varepsilon \qquad 2$$

in which ε_F is the Fermi energy for an electron in the electrode and $|V|$ is the magnitude of the electronic coupling between the redox species and the electrode. This description has been widely used in treating electron tunneling through organic thin films.

It has been found by many workers that the electron transfer rate

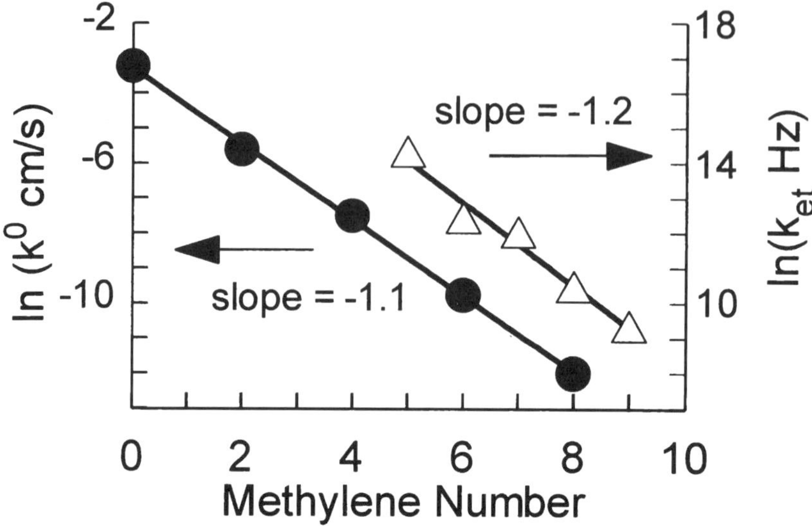

Figure 2: The logarithm of the rate constant is plotted as a function of the number of methylenes in the SAMs alkane chain. The circles are for $Fe(CN)_6^{3-/4-}$ and go with the left axis. The triangles are for ferrocene and go with the right axis. In both cases the solution is aqueous electrolyte with a viscosity around 1 cP. See text for further details.

constant decays exponentially with the thickness of the film. Figure 2 shows data from our work and that of Smalley [3] on Au electrodes. In each case the rate constant is plotted as a function of the number of methylene units in the alkane chain. Our data was taken for a redox couple, which can diffuse freely in solution whereas the Smalley data was obtained for a redox couple that is tethered to the electrode surface. In each case the rate constant's distance dependence is well characterized by an exponential fall off of about one per methylene unit. The distance dependence reflects the decrease in the electronic coupling with distance, which is well approximated by an exponential decay; *i.e.,*

$$|V| = |V_0| \cdot \exp\left(-\frac{\beta d}{2}\right)$$

3

When the electronic coupling between the electrode and the redox couple is large, the nonadiabatic picture is no longer appropriate and the electron transfer mechanism must be viewed differently. A number of workers have investigated the transition between the nonadiabatic and adiabatic regimes from a theoretical perspective [4], and Weaver [5] has examined it experimentally on bare electrodes. A unified expression for the electron transfer rate constant k_{et} is

$$\frac{1}{k_{et}} = \frac{1}{k_{et}^{NA}} + \frac{1}{k_{et}^{A}} \qquad 4$$

where k_{et}^{NA} is the nonadiabatic rate constant and k_{et}^{A} is the adiabatic rate constant. The adiabatic rate constant is given by

$$k_{et}^{A} = \nu_{eff} \cdot \sqrt{\frac{\lambda}{\pi^3 RT}} \cdot \exp\left(-\Delta G_{act}\middle/RT\right) \qquad 5$$

in which ΔG_{act} is the activation energy for the reaction, given by

$$\Delta G_{act} = \frac{(\lambda - \Delta_r G)^2}{4\lambda} - |V| \qquad 6$$

and $\Delta_r G$ is the reaction free energy. The parameter ν_{eff} characterizes the frequency of the medium's polarization response. This frequency may be characterized by the solution's characteristic solvation time [5,6] or, more approximately, its viscosity [2,6].

Figure 3 shows data that were obtained for electron transfer between the $Fe(CN)_6^{3-/4-}$ redox couple and a Au electrode as a function of the monolayer film thickness and the solution viscosity. For the bare electrode a viscosity dependence of the electron transfer rate constant is evident. As the monolayer thickness becomes thicker the electronic coupling between the electrode and the redox couple decreases, and the viscosity dependence of the electron transfer rate constant becomes weaker. For electrodes

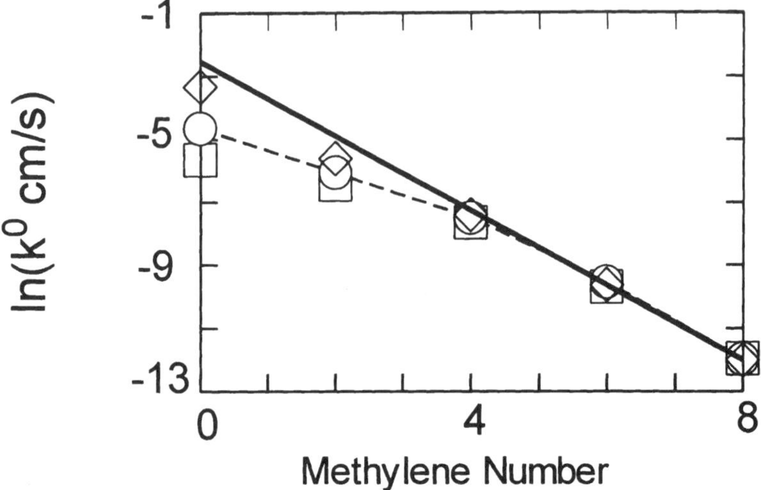

Figure 3: The logarithm of the rate constant is plotted against alkanethiol chain length at three different solution viscosities (squares are 11 cP, circles are 4 cP, and diamonds are 1 cP). The dashed curve connects the 4 cP data and is included as an aid to the eye. The solid line shows the rate dependence expected from the nonadiabatic picture. Adapted from Figure 3 of reference [2].

containing layers that have six or more methylenes, the rate constant does not depend on viscosity at all. In general, the viscosity dependence can be used as a signature for electron transfer in the strong coupling regime, whereas an exponential distance dependence for the electron transfer rate constant provides a signature of the nonadiabatic, or weak coupling, regime. These data and the transition between regimes is described in more detail elsewhere [2].

Composition Dependence of Tunneling Rate: The electronic coupling, which determines the tunneling probability, is commonly treated as occurring through a superexchange mechanism. If this view is appropriate then the electron transfer rate constant should display a dependence on the composition of the film. To explore the composition dependence and evaluate whether the electronic

Tethered Redox Couple

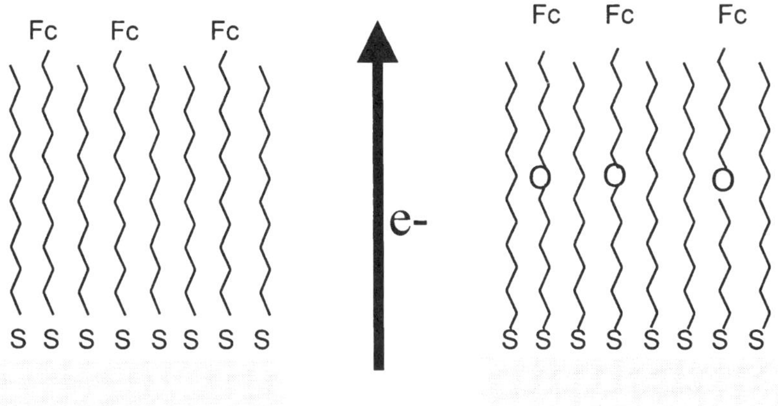

Figure 4: The schematic shows two of the redox active SAMs that were studied. The system on the left shows a ferrocene tethered to a gold electrode through an alkane link, and the system on the right shows a ferrocene tethered to a gold electrode through an ether link.

coupling was dominated by 'through bond' interactions [7], the electron transfer rate between a gold electrode and a tethered ferrocene was explored for four different systems. Figure 4 illustrates two of the systems that were investigated. In the system on the left a ferrocene moiety is tethered to a gold surface through an alkane chain, and in the system on the right it is tethered through an ether chain. In each of the cases shown the ferrocene coverage was *ca* 10% and the coverage of the alkane diluent thiol was 90%. In addition, the analogous two systems in which an ether-linked diluent was used instead of an alkane diluent were studied.

Table 1 presents the results obtained for the electron transfer rate constant of these four systems. The electron transfer rate constant is highest for the alkane-tethered ferrocene for both of the different diluent systems, ether and alkane. The rate constants for a

Table 1: Rate constants k_{et} and apparent formal potential, E^0, values are given for electron transfer through four different SAM films of similar thickness.

Ferrocene thiol	Diluent thiol	k_{et} (s^{-1})	E^0 (mV)	# of trials
Alkane	Alkane	55 ± 9	522 ± 13	10
Alkane	Ether	35 ± 12	532 ± 31	6
Ether	Alkane	12 ± 3	534 ± 13	3
Ether	Ether	8 ± 2	516 ± 32	7

given tether in the alkane and ether diluent are similar to each other, considering the experimental error. The electron transfer rate constant for the ether-linked ferrocene is significantly smaller (almost five times) than the alkane linked ferrocene. These data show the important role that the tether composition plays in changing the electronic coupling between the redox couple and the electrode and show that the coupling is dominated by through bond interactions for tethered systems. The similarity of the rate constants for the different diluent thiols and a given type of tether indicate that changes in the effective dielectric constant of the film have a small impact on the electronic transfer rate. Alternatively the small differences associated with the change of diluent thiol may reflect a small contribution from interchain coupling pathways [8]. A more detailed discussion of these results can be found elsewhere [9].

Quantum chemistry calculations were performed to identify the origin of the decrease in the electronic coupling. Calculations were performed at the 3-21G level on model diradical systems, namely $\cdot CH_2(CH_2)_{11}CH_2\cdot$ and $\cdot CH_2(CH_2)_5O(CH_2)_5CH_2\cdot$, for which different superexchange mechanisms, electron mediated versus hole mediated, can be evaluated. The electronic coupling magnitudes were obtained from splittings of the valence orbitals, assuming Koopmans approximation applies. The calculations show no difference in the electronic coupling for the electron mediated (radical anion; LUMO splittings) process, but they display a factor of two difference (alkane larger than ether) for the hole mediated (radical cation; HOMO splittings) process. In addition, the calculations show that the electronic coupling is largely dominated by non-nearest neighbor interactions and that destructive interference is an important component of the overall coupling. About 75% of the difference in coupling can be related to differences in the self energy of the CO bonds and the exchange interaction between the CO bonds and their nearest neighbors. These issues are discussed at length elsewhere [9].

The calculated electronic couplings are shown for different conditions in Table 2. After the calculation, the Fock matrix was transformed into the Natural Bond Orbital (NBO) basis to identify the mechanism of the superexchange from a more 'chemical' perspective. The total coupling is given in the first row of Table 2. If the core orbitals and the Rydberg orbitals of the chains are removed from the basis set, new electronic couplings are obtained that are significantly larger than that found for the complete basis. This results indicates that the core and Rydberg orbitals cause destructive interference and reduce the total coupling for both the ether and the alkane systems. If the σ orbitals are removed (only σ^* manifold remains) the coupling is very small, which shows that coupling through the σ^* manifold alone is not important. If the σ^* orbitals are removed (only σ manifold remains), the coupling is about half of that found with the full basis and shows that coupling

Table 2: Calculated electronic couplings are shown for different sets of basis orbitals.

NBO Calculation	alkane system $|V|_{NBO}$ (cm^{-1})	ether system $|V|_{NBO}$ (cm^{-1})
Full Basis Set	339	199
Full -Core	384	230
Full - Rydberg	372	211
σ and σ^*	419	242
σ manifold	158	86
σ^* manifold	1.2	1.4

through the σ manifold is important. However this manifold only represents about 40 to 50 percent of the total coupling. The dominant contribution to the coupling arises from processes that involve transitions between the σ and σ^* manifolds. Although such effects have been identified in earlier studies, they did not contribute a major portion of the coupling like that found here.

Electron Tunneling in Freely Diffusing Systems: A number of studies have indicated that interchain electronic coupling can play an important role in some systems [8,10]. In addition, it is found that the thickness dependence of the electron transfer rate constant between an electrode and a freely diffusing redox couple changes in a systematic manner between Hg, Au, and InP electrodes. The differences found in the observed thickness dependences appear to be correlated with the geometry of the alkane on the surface.

Recently Majda [8] studied the tilt angle dependence of the electron tunneling on alkanethiol coated Hg electrodes. They were able to systematically change the tilt angle of the alkane chain from the surface normal and found that the observed current changes with the tilt angle (see Figure 5). They proposed a simple two-pathway model to account for the observed dependence. One pathway is the 'through bond' superexchange along a chain (with a tunneling decay constant $\beta_{tb}=0.91$ Å^{-1}), and the second pathway involves an interchain process and has a significantly stronger distance dependence ($\beta_{ts}=1.31$ Å^{-1}). Figure 5 shows their data (squares) for the tunneling through a dodecanethiol monolayer films. The triangles in this figure are data found for gold electrodes [$\triangle$ from 1b; $\blacktriangle$ this work] and the filled circle is data for InP [11]. The dashed curve shows the predicted dependence for the model reported by Slowinski *et al* [8b], which only allows for a single interchain tunneling event. The solid curve shows the dependence predicted for the situation in which two interchain hops are allowed. A clear dependence on the tilt angle is evident for the model and can be used to rationalize the observed trend in the data. The electronic characteristics of these three electrode materials are different and other explanations for the behavior may be possible. For example, an increase in the electronic coupling between the redox couple and the electrode could cause a change in the electron transfer mechanism from the nonadiabatic case to an intermediate condition where the tunneling dependence would appear softer (*vide supra*). Clearly, more work is needed to verify this explanation for the observed tunneling dependence in an unequivocal manner. Nevertheless this result suggests that interchain coupling becomes more important for higher tilt angles.

These studies underscore important considerations in constructing electron tunneling barriers from organic molecules.

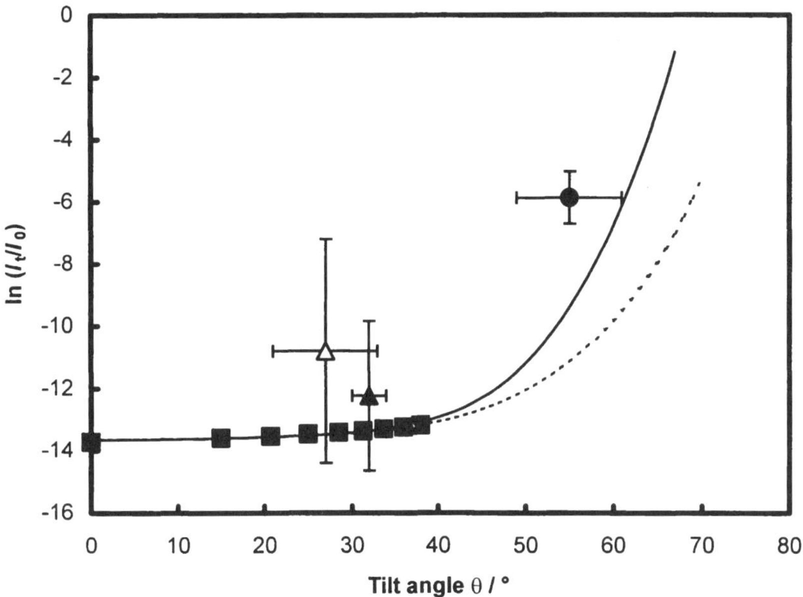

Figure 5. The normalized tunneling current is plotted as a function of the tilt angle for dodecanethiol self-assembled monolayers on Hg (■), InP (●), and Au ▲ (Δ is ω-hydroxydodecanethiol) electrodes. See text for further details. Figure is adapted from reference 11

This work shows that for covalently linked systems, the composition of the molecular unit that tethers the redox couple to the substrate plays an important role. The change in the electron transfer rate between the alkane and ether systems could be linked to changes in the molecule's electronic structure. The geometry of the alkane chains also plays an important role in the electron tunneling and it appears that interchain interactions become more important for more highly tilted alkane chains. However, more work on this issue will be required. Lastly, we showed that the mechanism of the electron transfer is a function of both the electronic coupling magnitude and the viscosity of the medium. When the mechanism changes from the nonadiabatic, or tunneling, regime to the adiabatic regime the electron transfer should become distance independent but display a dependence on the medium's polarization response. These effects may become manifest for transfer through highly conjugated systems [12].

Acknowledgements: We thank the Department of Energy (DE-FG02-89ER14062) and the NRC (through a Twinning Grant) for partial support of this work.

References

1 a) Finklea, H. O. *Electroanalytical Chemistry;* Bard, A. J., Rubinstein, I., eds.; Marcel Dekker: New York, **1996**; *Vol. 19*, 109; b) Miller, C. J. *Physical Electrochemistry: Principles, Methods and Applications*, Rubinstein I. ed. Marcel Dekker: New York, **1995**, 27.
2 Khoshtariya, D. E.; Dolidze, T. D.; Zusman, L. D. and Waldeck, D. H. *J. Phys. Chem. A* **2001**, *105*, 1818.
3 Smalley, J. F.; Feldberg, S. W.; Chidsey, C. E. D.; Linford, M.R.; Newton, M.D.; Liu, Y. P. *J. Phys. Chem.* **1995**, *99*, 13141.
4 a) Zusman, L. D. *Z. Phys. Chem.* **1994**, *186*,1; b) Alexandrov, I. V. *Chem Phys.* **1980**, *51*, 449; c) Hynes, J. T. *J. Phys. Chem.* **1986**, *90*, 3701; d) Heitele, H. *Angew. Chem. Int. Ed. Engl.* **1993**, *32*, 359; e) Sumi, H. *Chem. Phys.* **1996**, *212*, 9.
5 a) Weaver, M. J. *Chem Rev.* **1992**, *92*, 463; b) *ibid, J. Mol. Liqs.* **1994**, *60*, 57.

6   Maroncelli, M. *J. Mol. Liq.* **1993**, *57*, 1.

7   a) Newton, M. D. *Chem. Rev.* **1991**, *91*, 767; b) Jordan, K. D.; Paddon-Row, M. N. *Chem. Rev.* **1992**, *92*, 395.

8   a) Slowinski, K.; Chamberlain, R. V.; Majda, M.; Blewicz, R. *J.Am. Chem. Soc.* **1996**, *118*, 4709; b) Slowinski, K.; Chamberlain, R. V.; Miller, C. J. ; Majda, M. *J. Am. Chem. Soc.* **1997**, *119*, 11910.

9   Napper, A. M.; Liu, Haiying; Waldeck, D. H. J. Phys. Chem, **2001** *105*, 7699.

10 Finklea, H. O.; Liu, L.; Ravenscroft, M. S.; Punturi, S. *J. Phys. Chem.* **1996**, *100*, 18852.

11 Yamamoto, H.; Waldeck, D. H., *J. Phys. Chem.,* submitted.

12 Sikes, H. D.; Smalley, J. F.; Dudek, S. P.; Cook, A.R.; Newton, M. D.; Chidsey, C. E. D.; Feldberg, S.W. *Science* **2001**, *291*, 1519.

Chapter 7

Charge Transfer at Molecule–Metal Interfaces: Implication for Molecular Electronics

Gregory Dutton and X.-Y. Zhu[*]

Department of Chemistry, University of Minnesota, Minneapolis, MN 55455
[*]Corresponding author: zhu@chem.umn.edu

The operation of most conceivable electronic devices using organic molecules involves electron transfer at molecule-metal interfaces. Despite its prominent role, few studies in the molecular electronics community have addressed the fundamental issues, such as energetics and dynamics, involved in interfacial electron transport. Our laboratory has recently initiated a systematic study of this issue using laser two-photon photoemission spectroscopy. We demonstrate the important roles of molecule-metal wavefunction mixing, intermolecular band formation, and the presence of localized interfacial states in interfacial electron transfer.

Background

Building a successful molecular electronic device requires making electronic contacts to one or a group of molecules. In experiments, this is often the most difficult step. This difficulty arises because of the size mis-match between Å scale molecules and μm or sub-μm scale electrodes. In fact, some of the recent successes in demonstrating molecular electronic devices come from breakthroughs in making such contacts for electron transport.[1,2] From a theoretical perspective, while calculations on individual molecules and extended metal structures have both advanced significantly, treating the coupling of a localized molecular orbital to a delocalized metal band structure is no easy task.

The issues on electronic coupling and electron transfer across molecule-metal interfaces are not unique to molecular electronics. They are prevalent in many fields of chemistry and chemical technology. In organic light emitting devices and field-effect transistors, electron injection from metallic electrodes into the molecular layer is a critical step.[3-4] In surface photochemistry on metals, photo-induced charge transfer to molecular resonances is believed to be a dominant mechanism.[5] Interfacial electron transfer is also key to solar energy conversion and electrochemistry.[6]

In view of the fundamental importance of interfacial electron transfer in molecular electronics, our laboratory has initiated a systematic study using laser two-photon photoemission (2PPE) spectroscopy and model systems.[7,8,9,10,11] The 2PPE technique is uniquely suited to investigations on the energetics and dynamics of interfacial electron transfer.[12,13] Figure 1 shows the principles of the 2PPE technique in probing interfacial electron transfer. The molecular wavefunction (in the surface normal direction) shown schematically in the figure is characterized by a major component outside the metal surface and localized to the molecule, and a decaying "tail" inside the metal. The "tail" is a quantitative measure of the electronic interaction between the molecular state and the metal substrate. In the 2PPE process, the first photon excites an electron from an occupied metal state below the Fermi level to the unoccupied molecular resonance, and the second photon excites the electron from this transient state to a free-electron state above the vacuum level. The rates of both steps are given by Fermi's Golden Rule. For the first step, the rate is essentially proportional to the amplitude of the wavefunction inside the metal. The stronger the electronic interaction between the molecular resonance and the metal substrate, the larger the amplitude of the wavefunction inside the metal and the higher the rate of formation of the molecular anionic state. Note that in the above discussion, we assume that the first step in this 2PPE process corresponds to direct photo-excitation, rather than an indirect mechanism in which the molecular anionic state is formed from the scattering/relaxation of photo-excited substrate electrons into the molecular resonance. This direct photo-excitation mechanism has been verified by the dependence of 2PPE intensity on light polarization.[14] The evolution of the electronic wavefunction after the first photon absorption is followed in energy, momentum, and time domains by the second photon, thus providing unprecedented detail on such an interfacial electron-transfer event.

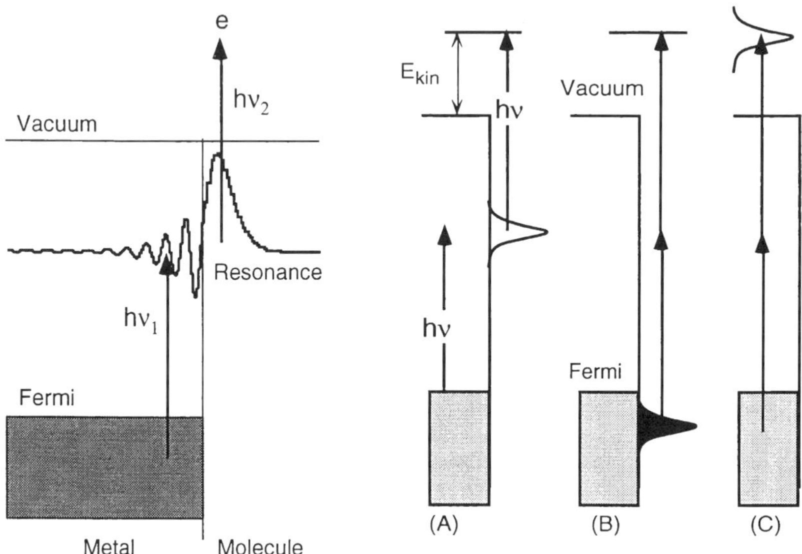

Figure 1. Schematic illustration of 2PPE probe of interfacial electron transfer involving an unoccupied molecular orbital.

Figure 2. Schematic illustration of three possibilities in 2PPE involving unoccupied states below (A) or above (C) vacuum, and an occupied state (B).

Two photon photo-emission probes both occupied and unoccupied states. Figure 2 shows schematically three possible scenarios in 2PPE. In (A), the absorption of the first photon excites an electron from occupied metal bands to an unoccupied intermediate state (above the Fermi level, below the vacuum level); the absorption of a second photon excites this transient electron above the vacuum level. In this case, the change in electron kinetic energy scales with that in photon energy, i.e., $\Delta E_{kin} = 1 \cdot \Delta h\nu$. In scheme (B), coherent two-photon excitation from a state below the Fermi level leads to the ejection of one electron. This is similar to conventional UPS, except that $\Delta E_{kin} = 2 \cdot \Delta h\nu$. Finally, if the unoccupied state is above the vacuum level, (C), E_{kin} is independent of photon energy. This can be viewed as a resonant scattering event in which the photoexcited electron resides transiently in the molecular resonance, followed by detachment and detection.

Two-photon photo-emission spectroscopy has been applied in the past to establish the dynamics of excited electrons in metal substrates, as well as image states on metal or dielectric covered metal surfaces.[12,13] This technique has also been used to establish unoccupied resonances at adsorbate-metal interfaces, including the π^* state of CO/Cu(111)[15] and σ^* state of Cs/Cu(111).[16] Our lab has extended this approach to interfacial electron transfer involving molecular resonances at organic-metal interfaces. We will show two model systems of significance to molecular electronics: (1) a weak interacting system, C_6F_6/Cu(111);[7,9,17] and (2) a strong chemisorption system of thiol self-assembled monolayers on Cu(111).[8] In the first system, electron transfer to the lowest unoccupied molecular orbital (LUMO) located at 3 eV above the Fermi level is observed in 2PPE. The effects of molecule-metal electronic coupling and intermolecular band formation on interfacial electron transfer are addressed. In the second system, we probe unoccupied electronic states in thiol SAMs on Cu(111). We find that interfacial electronic structure is dominated by two σ^*-like orbitals, independent of the nature of the hydrocarbon group (conjugated aromatics or saturated alkyls). The thiolate contact is insulating for electron transport through π^* LUMO in SAMs of molecular wires.

Role of molecule-metal interaction and intermolecular band formation in electron transfer: C_6F_6/Cu(111)

We choose this system because hexafluorobenzene, C_6F_6, has high electron affinity.[18,19] In the adsorbate state, the LUMO may be located significantly below typical positions for image states, thus facilitating interpretation. Figure 3 shows the thermal desorption spectra of C_6F_6 on clean and 0.34 ML atomic hydrogen covered Cu(111).[7,17] In both cases, the adsorption can be characterized as weak chemisorption. On clean Cu(111). C_6F_6 adsorbs in a stable bilayer structure on Cu(111) before multilayer formation. The monolayer desorbs at ~195 K with a shoulder feature at ~185K, and the second layer at 165 K. Atomic hydrogen is known to adsorb on bridge sites on the Cu(111) surface.[20] Ultraviolet photoemission showed that, for surface H prepared from atomic hydrogen adsorption on Cu(111), the surface state is removed (see below in 2PPE spectra); beyond this, its effect on substrate electronic structure is negligible.[21] The adsorption of H on Cu(111) weakens the chemical interaction between monolayer C_6F_6 and the Cu(111) surface, as evidenced by the up to 10 K decrease in the peak temperature. In fact, with increasing coverage of pre-adsorbed H (0-0.34 ML), there is a systematic shift in monolayer C_6F_6 desorption temperature.[17]

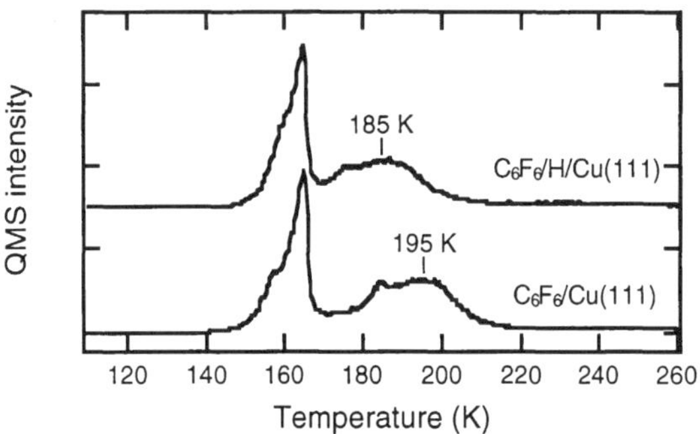

Figure 3. Thermal desorption spectra (TDS) of bilayer C6F6 on clean (bottom) and 0.34 ML H covered (top) Cu(111)

The left panel in figure 4 compares 2PPE spectra for clean Cu(111) and bilayer C_6F_6/Cu(111). The clean surface spectrum shows the intense surface state (SS) and the n = 1 image state (IS). The former originates from two photon photo-ionization of the occupied surface state while the latter involves one photon excitation into the n = 1 image state, followed by photoionization of this transient state. Also evident is the feature due to two-photon photoionization of the Cu d-band. With the adsorption of 2 ML of C_6F_6 on the Cu(111) surface, both the surface state and the image state are attenuated, and a new peak (MR for molecular resonance) at ~7 eV grows. The MR results from photoexcitation to an unoccupied state located at ~3 eV above the Fermi level, as evidenced by measurements with different photon energies (see below). This state is assigned to the σ^* LUMO of C_6F_6.

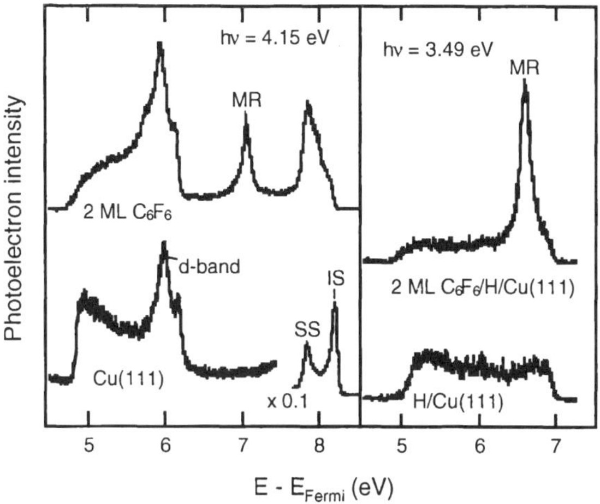

Figure 4. 2PPE spectra taken at two photon energies: hv = 4.15 eV (left) and 3.49 eV (right). Left panel: clean Cu(111) & 2 ML C_6F_6/Cu(111); Right panel: 0.30 ML H/Cu(111), and 2 ML C_6F_6 on 0.30 ML H/Cu(111). SS = surface state; IS = image state; MR = molecular resonance.

The right panel in figure 4 shows 2PPE spectra taken at hv = 3.49 eV for 0.30 ML H/Cu(111), and bilayer C_6F_6 on 0.30 ML H/Cu(111). The adsorption of H on Cu(111) eliminates the surface state and the further adsorption of bilayer C_6F_6 on H/Cu(111) leads to a well defined resonance (MR). The unoccupied nature of the resonance is established by the dependence of 2PPE peak position on photon energy, shown in figure 5 for bilayer C_6F_6 on 0.34 ML H/Cu(111). Similar results are obtained for bilayer C_6F_6 on clean Cu(111).[7] The data can be well described by a straight line, with a slope of n = 1.09 ± 0.10, which is essentially unity. According to figure 2(A), the observed resonance for bilayer C_6F_6 on H covered Cu(111) originates from photo-excitation to an unoccupied state located at 3.22 eV above the Fermi level, followed by photo-ionization of this transient state. This state is again assigned to the σ^* LUMO of C_6F_6.

Compared to the spectrum on clean Cu(111), the presence of 0.34 ML atomic H shifts the molecular resonance in C_6F_6 upward by 0.24 eV. In addition, the intensity of the MR peak is decreased by nearly one-order of magnitude. Both effects by H can be attributed to the weakening of molecule-metal interaction, as illustrated in figure 1. Such a weakening in electronic coupling between the two is quantified as a decrease in the amplitude of wavefunction (tail) inside the metal substrate. This results in a decrease in the rate of photo-induced electron transfer from the metal substrate to the molecular resonance. The mixing of molecular wavefunction with substrate bands is also partially responsible for the stabilization of the transient anionic molecular resonance on the metal surface. Decreasing this mixing by the pre-adsorption of H will reduce the effect of stabilization, thus, shifting the MR upward in energy.

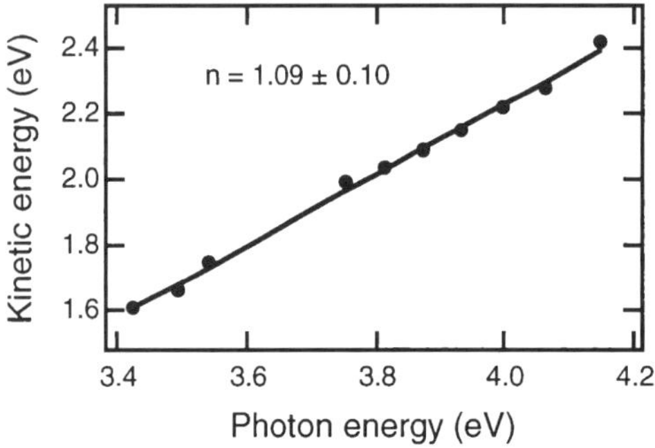

Figure 5. The dependence of the kinetic energy of photoelectrons from the molecular resonance as a function of photon energy for bilayer C_6F_6 on 0.34 ML H-covered Cu(111).

The molecular resonance in C_6F_6 on clean and H-covered Cu(111) displays quantum-well behavior, i.e., delocalization parallel to the surface. Figure 6 shows dispersion curves obtained from angle-resolved measurements. On both clean and 0.34 ML H covered Cu(111), the molecular resonance in C_6F_6 is dispersed parallel to the surface. The solid lines are parabolic fits to free electron like dispersion. The fits yield the following effective electron mass: $m_{eff} = 1.5\ m_e$ and $0.55\ m_e$ (m_e: free electron mass) on clean and H-covered Cu(111), respectively. We believe the observed free electron like behavior parallel to the surface results from inter-molecular electronic band formation. With increasing coverage of pre-adsorbed H, which weakens the molecule-surface interaction, the intermolecular wavefunction overlap is improved for electronic band formation due to the improved molecule-molecule packing. Thus, the effective electron mass for C_6F_6 on H covered Cu(111) is less than that on clean Cu(111).

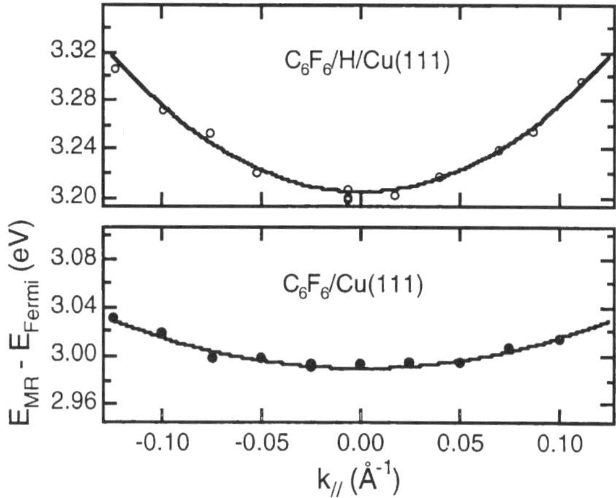

Figure 6. Parallel dispersion curves for the molecular resonance in bilayer C_6F_6 on clean Cu(111) (lower panel) and 0.34 ML H/Cu(111) (upper panel). The solid curves are fits to the free electron like parabolic relationship.

Interfacial states in thiol self-assembled monolayers and their role in electron transfer through molecular wires

Self-assembled monolayers (SAMs) of thiols on metal surfaces have been popular choices in the construction and testing of molecular electronic devices. The easy formation of thiolate-metal contact is an attractive method to connect molecular components to metal electrodes. The questions are: What is the electronic structure of the thiolate-metal contact? How does this contact affect electron transport through the molecular wire? To address these questions, we carried out 2PPE measurement in conjunction with ab initio calculation on model SAM/Cu(111) systems.[8]

We choose SAM of thiophenolate on Cu(111) as a model for molecular "wire" - metal electrode interfaces. The left panel in figure 7 shows a set of 2PPE spectra for a SAM of thiophenolate, C_6H_5-S-, on Cu(111) as a function of photon energy. There are two resonances: the position of the first (A) stays constant while that of the second shifts with $1\Delta h\nu$. Based on figure 2, the two resonances in figure 7 (left) must correspond to electron transfer to a final state (above the vacuum level) and an intermediate state (between the Fermi and the vacuum level), respectively. These two states are located at 6.4 eV and 3.3 eV above the Fermi level, respectively. Nearly identical results are obtained for SAMs of saturated alkanethiolates, such as ethylthiolates on Cu(111) in the right panel. The spectra again reveal two unoccupied molecular resonances at nearly the same positions as those on the thiophenolate covered surface.

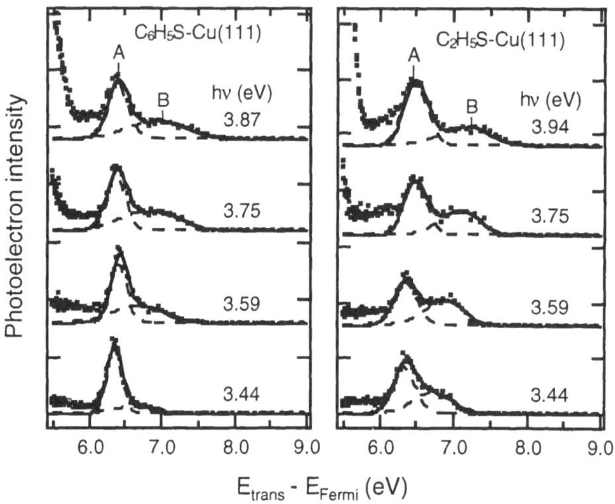

Figure 7. 2PPE spectra for SAMs of thiophenolate (left) and ethylthiolate (right) on Cu(111). Each spectrum was taken at the indicated photon energy.

These results suggest that the observed molecular resonances do not originate from the molecular framework, but rather from a common unit: the C-S-Cu linker. This hypothesis is supported by ab initio calculation on Cu-thiolate molecules. As shown in figure 8, the lowest unoccupied molecular orbitals (LUMO) for Cu-thiophenolate (Cu-S-C_6H_5) and Cu-propylthiolate (Cu-S-C_3H_7) are both σ^* states localized to the Cu-S bond, while the LUMO+1 state, located at 2-3 eV above the LUMO, is a combination of σ^* antibonding localized to S-C and π bonding between S and Cu.

For symmetry reasons, these localized σ^* states introduced by the anchoring bond cannot couple to the delocalized π^* states within the conjugated molecular framework. Thus thiolate contact can be considered "insulating" for electron transport through a self-assembled monolayer of molecular wires. On the other hand, ab initio calculation shows that the occupied molecular orbital (HOMO) is the π orbital delocalized between the molecular framework and the metal surface via the –S– bridge. Therefore, unlike for electron transport, the thiolate contact is conducting for hole transport.

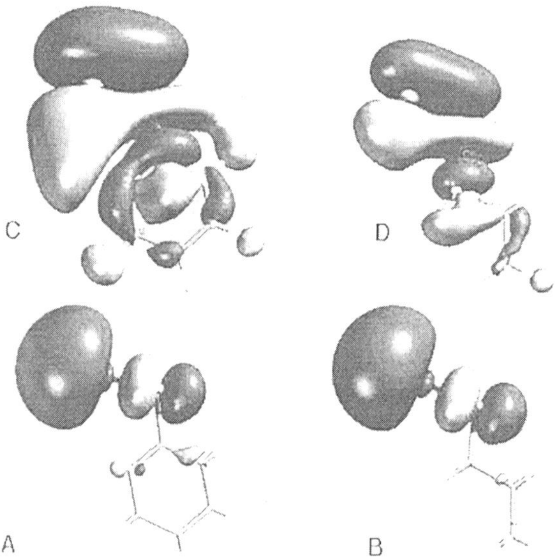

Figure 8. A & C: LUMO and LUMO+1 molecular orbitals for copper(I) thiophenolate; B & D: LUMO and LUMO+1 molecular orbitals for copper(I) propanethiolate.

Acknowledgement:

Financial support came from the National Science Foundation, grant DMR 9982109. Acknowledgement is made to the donors of the Petroleum Research Fund, administered by the ACS, for partial support of this work.

References

[1] Reed, M. A.; Zhou, C.; Muller, C. J.; Burgin, T. P.; Tour, J. M. *Science* **1997**, *278*, 252.

[2] Chen, J.; Reed, M. A.; Rawlett, A. M., Tour, J. M. *Sciecne*, **2000**, *286*, 1550.

[3] Garnier, F.; Kouki, F.; Hajlaoui, R.; Horowitz, G. *MRS Bulletin*, **1997**, *22*, 52.

[4] Salaneck, W.R.; Brédas, J.L. *MRS Bulletin*, **1997**, *22*, 47.

[5] Zhu, X.-Y. *Annu. Rev. Phys. Chem.* **1994**, *45*, 113.

[6] Miller, R. J. D.; McLendon, G. L.; Nozik, A. J.; Schmickler, W.; Willig, F. *Surface Electron Tansfer Processes* (VCH: New York, **1995**).

[7] Vondrak, T.; Zhu, X.-Y. *J. Phys. Chem. B* **1999**, *103*, 3449.

[8] Vondrak, T.; Wang, H.; Winget, P.; Cramer, C. J.; Zhu, X.-Y. *J. Am. Chem. Soc.* **2000**, *122*, 4700.

[9] Zhu, X.-Y.; Vondrak, T.; Wang,, H; Gahl, C.; Ishioka, K.; Wolf, M. *Surf. Sci.* **2000**, *451*, 244.

[10] Wang, H.; Dutton, G.; Zhu, X.-Y. *J. Phys. Chem. B*, **2000**, *104*, 10332.

[11] Vondrak, T.; Cramer, C. J.; Zhu, X.-Y. *J. Phys. Chem. B* **1999**, *103*, 8915.

[12] Fauster, Th.; Steinmann, W. in *Electromagnetic Waves: Recent Developments in Research*; Halevi, P., Ed.; (Elsevier: Amsterdam, **1995**).

[13] Harris, C. B.; Ge, N.-H.; Lingle, R. L; McNeill, J. D.; Wong, C. M. *Annu. Rev. Phys. Chem.* **1997**, *48*, 711.

[14] Wolf, M.; Hotzel, A.; Knoesel; E.; Velic, D. *Phys. Rev. B*, **1999**, *59*, 5926;

[15] Hertel, T.; Knoesel, E.; Hasselbrink, E.; Wolf, M.; Ertl, G. *Surf. Sci.* **1994**, *317*, L1147

[16] Petek, H.; Weide, M. J.; Nagano, H.; Ogawa, S. *Science* **2000**, *288*, 1402.

[17] Dutton, G.; Zhu, X.-Y. *J. Phys. Chem.* submitted for publication.

[18] Christophorou, L. G.; Datskos, P. G.; Faidas, H. *J. Chem. Phys.* **1994**, *101*, 6728.

[19] Dudde, R.; Reihl, B.; Otto, A. *J. Chem. Phys.* **1990**, *92*, 3930.

[20] McCash, E. M.; Parker, S. F.; Pritchard, J.; Chesters, M. A. *Surf. Sci.* **1989**, *215*, 363.

[21] Greuter, F.; Plummer, E. W. *Solid State Comm.* **1983**, *48*, 37.

Materials

Layered Nanoparticle Architectures on Surfaces for Sensing and Electronic Functions

Itamar Willner, Andrew N. Shipway, and Bilha Willner

Institute of Chemistry, The Farkas Center for Light-Induced Processes,
The Hebrew University of Jerusalem, Jerusalem 91904, Israel

Nanoparticle architectures on surfaces reveal unique sensoric and electronic functions. For instance, receptor-crosslinked Au-nanoparticles on electrodes act as sensing interfaces by the concentration of analytes in receptor sites. The 3D conductivity of the layered arrays enables the electrochemical detection of substrates, and the sensitivity of the system is controlled by the number of nanoparticle layers. Layered structures of DNA-functionalized CdS-nanoparticles allow the optical and photoelectrochemical detection of DNA. The fluorescence and photocurrents resulting from the systems are controlled by the number of aggregated nanoparticle layers. Finally, the electroless deposition of gold on avidin-Au-nanoparticle conjugates provides a means for the amplified microgravimetric detection of DNA.

Increasing interest is directed to the miniaturization and nanoscale engineering of functional chemical assemblies.[1,2] Metal and semiconductor nanoparticles exhibit properties that bridge molecular and bulk properties, and they thus provide important building blocks for engineering on this scale.[3,4] Nanoparticles can be synthesized from a variety of materials with controllable sizes, shapes and morphologies.[5] Their electronic properties, which can be tuned by the control of the nanoparticle dimensions, give rise to unique absorbance and emission features,[6,7] including plasmon excitons.[8] Nanoparticles also offer catalytic abilities that differ from the bulk material properties due to their high surface area and edge concentration.[9]

The surface modification of nanoparticles provides a means to stabilize them against reaction or precipitation, to link identical or different nanoparticles to each other, and to assemble them on surfaces.[10] The engineering of nanoparticles by chemical and self-assembly techniques has led to the construction of nanoparticle dimers,[11] trimers,[12] controlled aggregates,[13] wires,[14] ordered monolayers[15] and multilayers.[16] The integration of chemically functionalized nanoparticles with surfaces has also enabled the organization of nanoscale devices, including single-electron devices.[17,18] Here, we will describe the assembly of three-dimensional structures of metal or semiconductor nanoparticles on electrodes. The sensoric, electronic, optical and photoelectrochemical functions of these architectures will be addressed.

Receptor-crosslinked Au-nanoparticle architectures for sensoric applications

Scheme 1 outlines a method for the construction of three-dimensional (3D) receptor-crosslinked Au-nanoparticle architectures by layer-by-layer deposition.[19] A conductive ITO glass support is functionalized by a thin film of polymerized 3-(aminopropyl)triethoxysilane, generating a positively charged interface. Negatively charged citrate-capped Au-nanoparticles (12$\square$1 nm diameter) are adsorbed onto this surface, and the assembly is then interacted with the tetracationic cyclophane *cyclobis*(paraquat-*p*-phenylene) (**1**), which electrostatically binds to the Au-nanoparticles. By a stepwise layer-by-layer deposition process, a molecular-crosslinked nanoparticle structure of a controllable thickness is constructed.

Figure 1(A) shows the increase in the plasmon absorbance of the Au-nanoparticles upon the build-up of the layers. In addition to the increase in the plasmon absorbance at 520 nm, a second absorbance band appears at ≈650 nm, which intensifies as the number of Au-nanoparticle layers increses. This absorbance is attributed to interparticle plasmon coupling brought about by the aggregation of the nanoparticles on the surface. Figure 1(B) shows cyclic voltammograms of the Au surface of 1-5 layer assemblies (in 1.0 M H$_2$SO$_4$). Coulometric analysis of the reduction waves indicates an almost linear increase in the gold content with the number of layers, and an average coverage of ca. 1x10^{13} particles·cm^{-2} per layer. Figure 1(C) shows cyclic voltammograms of the tetracationic cyclophane crosslinker **1** upon the build-up of the assembly. Coulometric analysis of the redox-waves indicates a linear increase in the electrochemical response of **1** upon the construction of the layers, with an average surface coverage of ca. 1.5x10^{-11} mol cm^{-2} per layer. Thus, about 100 molecules of **1** are associated with each nanoparticle in the array. The linear increase in the redox-response of **1** upon the assembly of the crosslinked structure also implies that the architecture exhibits 3D conductivity and porosity.

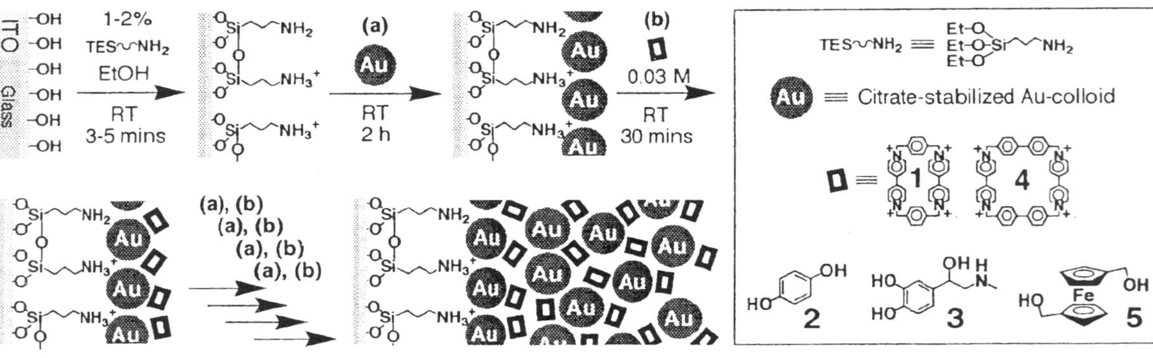

Figure 1. (A) Absorbance spectra of glass slides bearing 1-5 *1*-crosslinked Au-nanoparticle 'layers'. (B) Cyclic voltammograms of the Au-surfaces, and (C) cyclic voltammograms of *1*, in identical assemblies to those in (A).

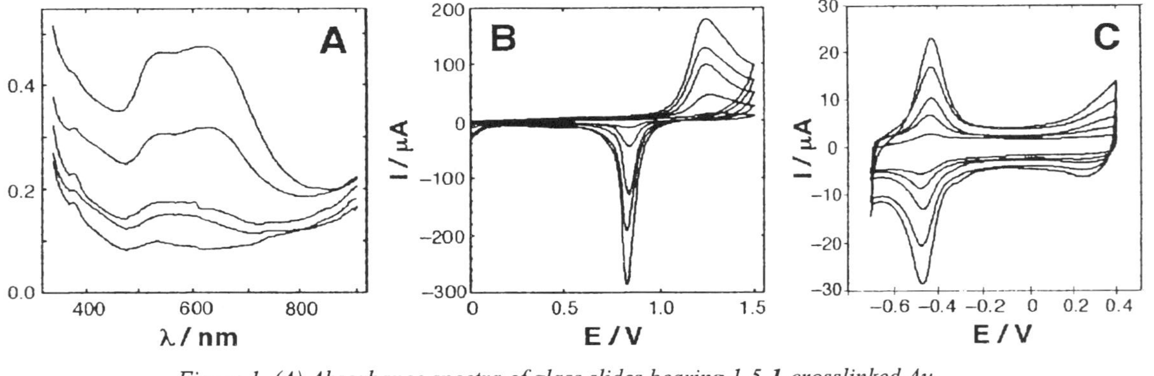

Scheme 1. The construction of oligocation-Au-nanoparticle composites on glass surfaces by a stepwise electrostatic approach.

The crosslinking unit **1** acts as a π-acceptor receptor and is capable of forming π-donor-acceptor complexes with various substrates.[20] The association of an electroactive substrate to the recognition sites, together with the 3D conductivity of the architecture therefore enables its electrochemical detection. Figure 2(A) shows the electrochemical analysis of *p*-hydroquinone (**2**) by a five 'layer' **1**-crosslinked Au-nanoparticle functionalized electrode. The sensitivity for the sensing of **2** is controlled by the number of crosslinked Au-nanoparticle layers on the electrode. A five layer Au-nanoparticle architecture crosslinked by *N,N'*-dimethyl-4,4'-bipyridinium does not respond to **2** in this concentration range. This indicates that the electrochemical sensing of **2** does not originate from the increase in the surface-roughness of the electrode, but rather from the concentration of **1** by interaction with the receptor units. The 3D-conductivity of the array enables the electrochemical detection of **2**. Other π-donor substrates can also be electrochemically sensed by this engineered nanoparticle structure.[21] Figure 2(B) exemplifies the electrochemical detection of the neurotransmitter adrenaline (**3**).

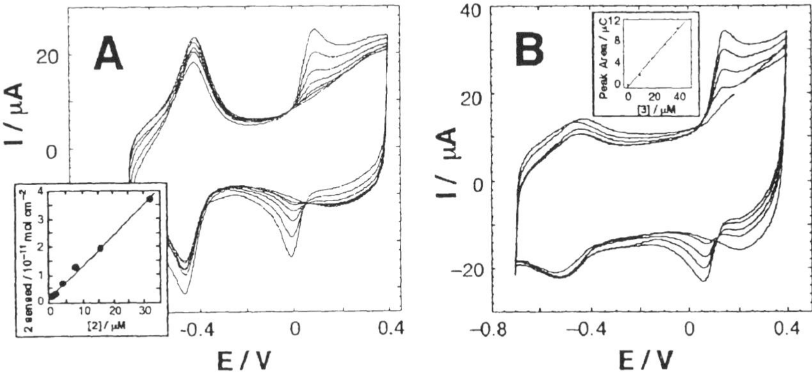

Figure 2. (A) The sensing of hydroquinone (2) (1-30 μM) by a 5-layer 1-crosslinked Au-nanoparticle array. Inset: Calibration curve. (B) The sensing of adrenaline (3) (1-40 μM) by the array. Inset: Calibration curve.

The specificity of sensing by the 3D-Au-nanoparticle array can be changed by altering the structure of the crosslinking receptor units.[22] An array of Au-nanoparticles was constructed using the enlarged cyclophane *cyclobis*(paraquat-*p*-biphenylene) (**4**) (Scheme 1). Figure 3(A) shows the increase in the plasmon absorbance band and the appearance of the interparticle plasmon absorbance upon the build-up of the layers and aggregation of the nanoparticles. Figures 3(B) and (C) show cyclic voltammograms of the Au surface (in 1.0 M H$_2$SO$_4$) and **4** upon the build-up of layers. Coulometric analyses indicates a linear increase in the surface coverage of the nanoparticles and **4**. An average of ca. 50 molecules of **4** are associated with each particle.

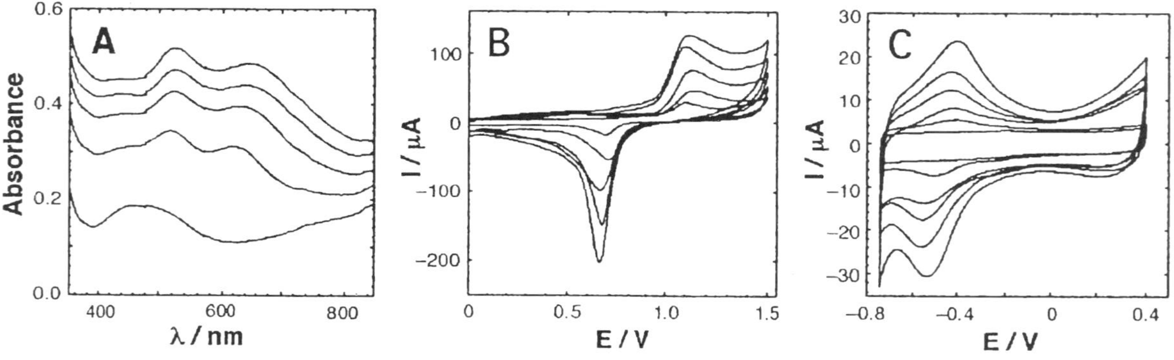

Figure 3. (A) Absorbance spectra of slides bearing 1-5 4-crosslinked Au-nanoparticle 'layers' on a glass substrate. (B) Cyclic voltammograms of the Au-surface of 1-5 'layer' 4-crosslinked assemblies of Au-nanoparticles. (C) Cyclic voltammograms of 4 in 1-5 'layer' arrays of 4-crosslinked Au-nanoparticles.

The enlarged π-acceptor receptor unit can accommodate *bis*-hydroxymethylferrocene (**5**) as a π-donor guest substrate. Figure 4(A) shows the analysis of **5** (1×10^{-6} M) by **4**-crosslinked Au-nanoparticle electrodes with different numbers of crosslinked Au-nanoparticle layers. As the number of layers increases, so does the electrical response of **5**, demonstrating tunable sensitivity of the nanostructured electrode. Figure 4(B) shows the amperometric responses of the Au-nanoparticle array upon exposure to different concentrations of **5**. The arrays also reveal selectivity. While **5** is electrochemically sensed by the **4**-crosslinked Au-nanoparticle structure, this receptor is too large to accommodate *p*-hydroquinone (**2**) in its cavity. Similarly, the **1**-crosslinked Au-nanoparticle assembly is unable to accommodate **5**, but does associate with *p*-hydroquinone (Figure 4(C)).

DNA-crosslinked nanoparticle arrays

Double-stranded DNA provides a different building unit for crosslinking nanoparticles on surfaces, and several advantages of this motif should be emphasized: (i) The base-complementarity in double-stranded DNA yields rigid spacing units of controlled length and width. (ii) Complex DNA structures can be synthesized,[23] e.g. hairpins, triangles, cubes. A battery of enzymes is available for nucleic acid elongation (ligase), polymerization (polymerase) and specific scission (endonuclease), giving rise to the biocatalytic control of the spacer length. (iii) Different nanoparticles, functionalized with different nucleic acids complementary to the two ends of a common nucleic acid, may be hybridized to yield arrays of crosslinked nanoparticles of variable sizes or composition. (iv) Double-stranded DNA crosslinking units may be used as a template for the chemical deposition of conductive wires,[24] thus enabling the construction of complex nanoscale circuitry. Exciting applications and devices are expected to emerge from basic knowledge of DNA-crosslinked nanoparticle architectures, and the field of DNA-bioelectronics could expand from biosensing devices to DNA computers and memory devices. The unique optical, electronic and photoelectrochemical functions of the nanoparticles may be used to probe and analyze DNA. Charging of particles by single-electron processes, the application of the DNA bridges as resonance tunneling units, and the photonic activation of semiconductor particles, may ultimately lead to dense memory devices and switching elements.

Bridging of Au-nanoparticles in solution has been used as a general strategy to develop colorimetric DNA detection. The shift of the characteristic red nucleic-acid-functionalized Au-nanoparticle plasmon absorbance to the blue color characteristic of coupled plasmons, in the presence of an analyte DNA that aggregates the nanoparticles has been used to develop various DNA detection routes.[25] The crosslinking of metal or semiconductor nanoparticles on surfaces was employed recently for DNA detection.[26,27] Scheme 2 outlines the assembly of layered Au-nanoparticle arrays on Au/quartz piezoelectric crystals or on glass supports. The thiolated primer (**6**) was assembled on the Au/quartz crystal or reacted with a maleimide-functionalized glass surface. This primer was reacted with the target DNA, (**7**), to yield double-stranded assemblies on the respective surfaces.

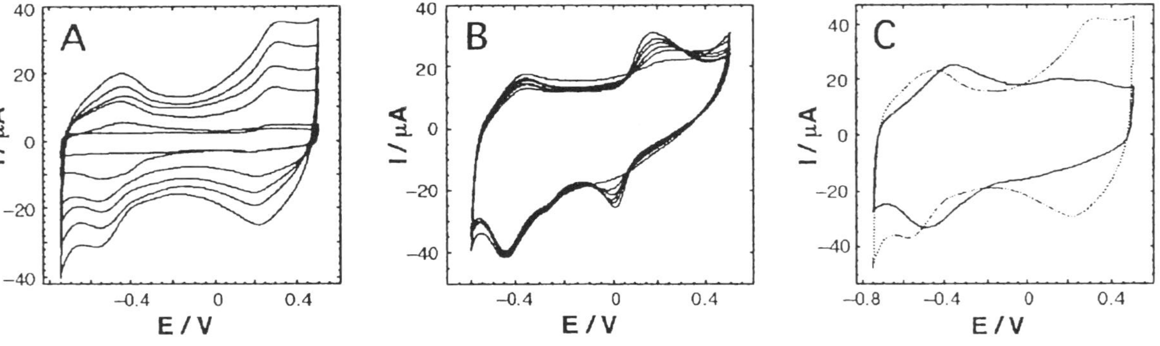

Figure 4. (A) Cyclic voltammograms for the sensing of 5 (1 µM) with 0-5 layer 4-crosslinked architectures. (B) The sensing of 5 (0.1-3 µM) by a 5-layer 2-crosslinked array. (C) Cyclic voltammograms of 5-layer electrodes constructed using 1 (unbroken line) and 2 (dotted line), in the presence of 6 (1 µM).

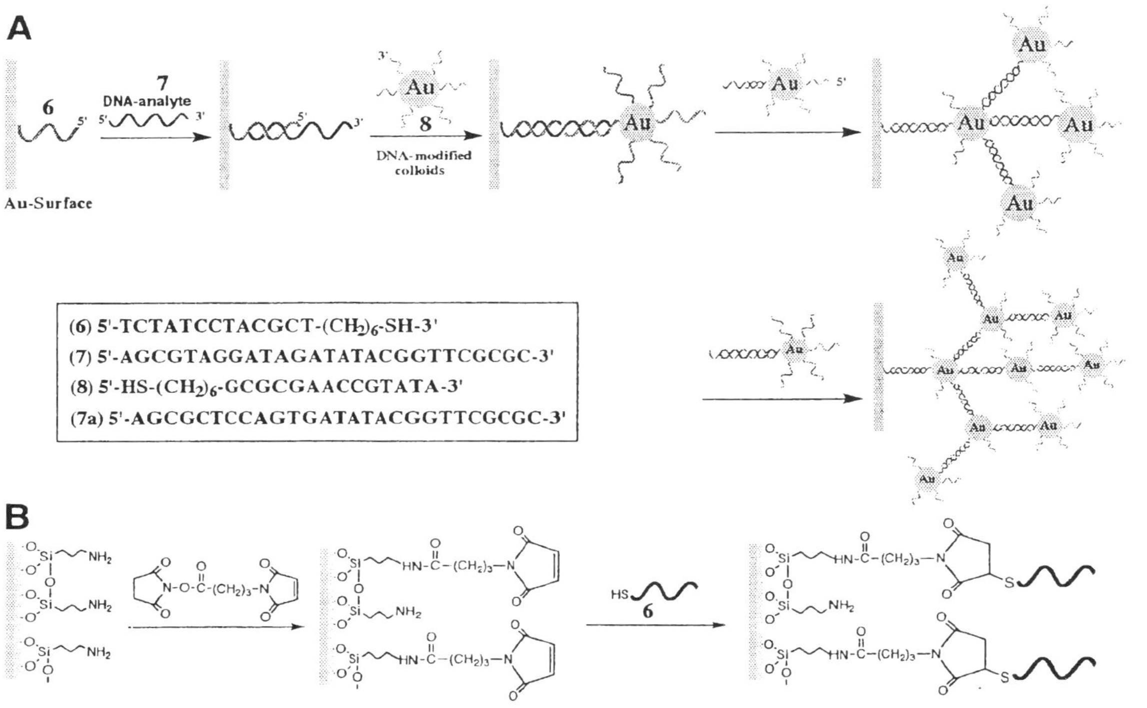

Scheme 2. (A) The sensing of a DNA analyte by the association of Au-nanoparticles to a Au-surface, and the 'dendritic-type' amplification of the assembly. (B) A methodology for the linking of DNA to a glass surface.

For the construction of the Au-nanoparticle array, Au-nanoparticles functionalized with nucleic acids complementary to the two ends of the target DNA were prepared: One type consists of the (**8**)-functionalized Au-nanoparticle, whereas the second class of nanoparticles consists of (**6**)-functionalized Au-nanoparticles. Interaction of the (**6**)/(**7**) double-stranded assembly with the (**8**)-functionalized Au-nanoparticles generates the first layer of particles. Reaction of the first layer of Au-nanoparticles with the (**6**)-functionalized Au-nanoparticles, that are pre-hybridized with (**7**), generates the second layer of nanoparticles. Thus, using the two types of modified nanoparticles enables the construction of a structure with a controllable number of layered particle aggregates. The association of the Au-nanoparticle to the piezoelectric quartz-crystal alters its mass, and the resulting frequency changes provide an electronic transduction for the primary association of the target DNA (**7**) with the surface. The stepwise construction of the layered array enables the dendritic amplification of the detection of (**7**).[26]

A similar method was employed to construct DNA-crosslinked CdS-nanoparticle assemblies.[27] Figure 5 shows the frequency changes of an Au/quartz crystal on which various 'generations' of DNA-crosslinked CdS aggregates are assembled. A non-linear variation in the frequency changes is observed upon the build-up of the layers, indicating a dendritic-type aggregation of the CdS-nanoparticles.

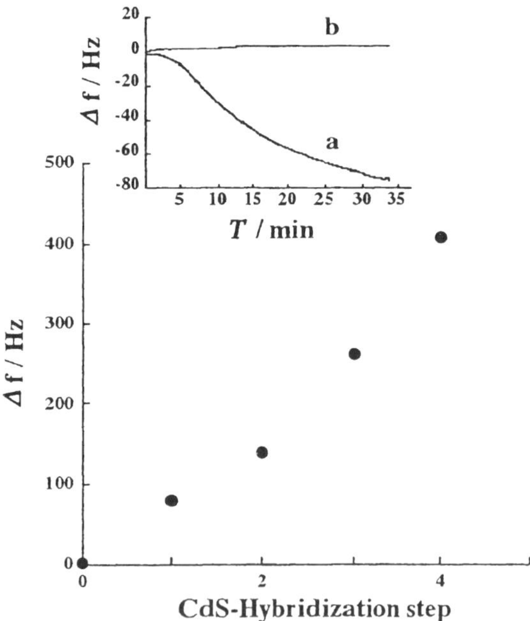

Figure 5. Frequency changes of a quartz crystal upon the construction of a 'dendritic-type' CdS architecture (0-4 layers). Inset: Kinetics of the association of the first layer of CdS (a) using the complementary anayle 7 and (b) using the non-complementary anayle 7a.

The DNA crosslinked architectures of CdS particles were also assembled on glass supports. The absorbance and fluorescence spectra of the CdS-nanoparticle layered aggregates are depicted in Figure 6. The non-linear increase in the absorption and fluorescence bands upon the construction of the shell-assemblies further supports the dendritic features of the nanoparticle aggregates. Also, the large stock-shift in the fluorescence spectra suggests that the emission originates from surface traps. This is consistent with the experimental procedure utilized to prepare the CdS nanoparticles that favor the formation of a Cd^{2+}-rich surface, that facilitates the subsequent binding of the thiolated oligonucleotide capping layer. These Cd^{2+}-sites act also as surface traps for conduction-band electrons.

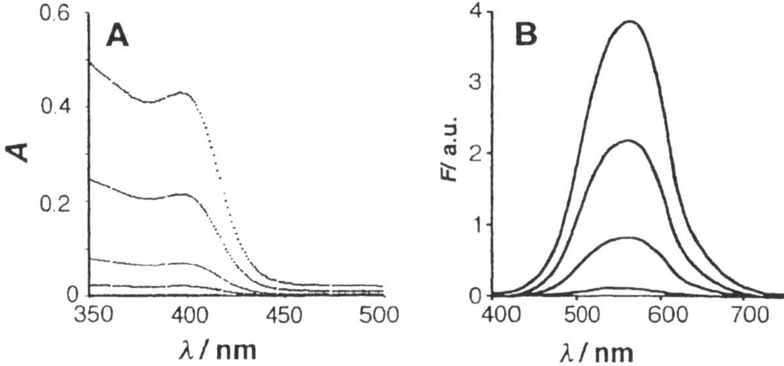

Figure 6. (A) Absorbance and (B) flourescence spectra of 1-4 layer 'dendritic-type' CdS superstructures.

The layered CdS-nanoparticle structures act as active photoelectrochemical interfaces. Figure 7 shows the photocurrent action spectra of the DNA-crosslinked CdS arrays, consisting of different numbers of aggregated layers. The photocurrent spectra follow the absorption spectrum of the CdS-particles (see inset, Figure 7). The photocurrents increase as the number of aggregated CdS layers are elevated. The mechanism for the generation of the photocurrents by the nanoparticle aggregates is, by itself, an interesting issue. The DNA units linking the CdS nanoparticles are non-conductive bridges[28]. Thus, the observed photocurrents imply that the charge-injection into the electrode originates from CdS nanoparticles that are in intimate contact, or at a direct electron tunneling (or hopping) distance, in respect to the conductive support. This might suggest that a significant portion of the CdS particles in the aggregates are inactive in the generation of the observed photocurrent. Indeed, the photocurrents are enhanced upon the addition of $Ru(NH_3)_6^{3+}$ to the system.[27] The $Ru(NH_3)_6^{3+}$ complex associates to DNA[29] and its reduction potential, $E = -0.16$ V vs. SCE, allows the transfer of photoexcited conduction-band electrons to the transition metal complex associated with the DNA. Thus, the $Ru(NH_3)_6^{3+}$ acts as an electron-relay, and the transport of electrons through the relay units facilitates the generation of the enhanced photocurrent.

98

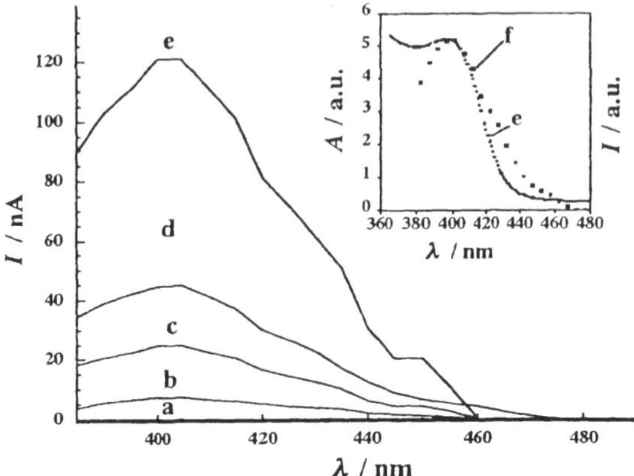

Figure 7. Photocurrent spectra of 0-4 layer 'dendritic-type' CdS arrays (a-e).
Inset: comparison of the absorbance (e) and photocurrent (f) spectra.

The formation of the photocurrent by the layered CdS-nanoparticle aggregates enables the sensing of an analyte DNA. Figure 8 shows the photocurrents generated by a 2-layer DNA-CdS crosslinked aggregate upon the analysis of different concentrations of (7). In this experiment a 2-layer aggregate is generated in order to utilize the dendritic amplification features of the aggregate upon sensing of (7). The aggregate on the electrode surface is then treated with the analyte samples that include different concentrations of (7)

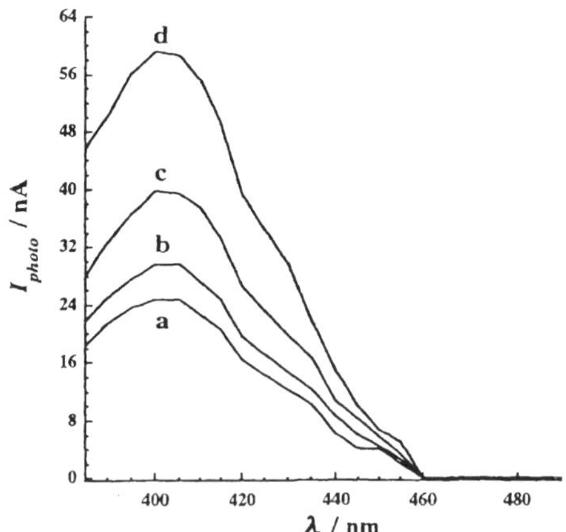

Figure 8. Photocurrent spectra of 2 layer (a and c) and 4 layer (b and d) CdS
arrays in the absence (a and b) and presence (c and d) of $Ru(NH_3)_6^{3+}$.

which were pretreated with the (**8**)-functionalized CdS-nanoparticles. A 3-layered generation of CdS-nanoparticles is formed resulting in an increase in the photocurrents. As the concentration of the analyte DNA (**7**) increases, the surface coverage of the third layer of CdS nanoparticles increases, resulting in enhanced photocurrents.

The formation of the layered CdS/DNA crosslinked aggregate, and the resulting photocurrent in the system, has important implications beyond the electronic detection of DNA. Our studies describe the development of programmed architectures of metal or semiconductor nanoparticles crosslinked by DNA bridging units. The stepwise assembly of the structures suggests that metal and semiconductor nanoparticle composite arrays may be envisaged. Also, the DNA bridging units may act as templates for the deposition of other materials, e.g. metals, or as host units for the intercalation of guest substrates. Complex material circuitries and tailored matrices for electron transport are expected to originate from such studies.

Amplified microgravimetric DNA detection by Au-nanoparticle-catalyzed electroless deposition of gold

The catalytic properties of nanoparticles may be further utilized to develop amplified detection schemes for DNA. The enzyme-catalyzed precipitation of an insoluble product on electronic transducers was employed to develop different amplified detection processes for the analysis of DNA[30] or antibodies.[31] The association of an enzyme conjugate to the double-stranded DNA assembly or the antigen-antibody complex, and the subsequent biocatalyzed precipitation of a product provides an amplification route since a single recognition event leads to the formation of many insoluble precipitate molecules. The biocatalyzed precipitation of an insoluble organic substrate on electrodes insulates the conductive support, and the changes in the electron transfer features at the electrode interface could be followed by electrochemical means, e.g. Faradaic impedance spectroscopy or chronopotentiometry. Alternatively, the formation of the precipitate on a piezoelectric quartz crystal changes the crystal frequency as a result of mass accumulation on the crystal. Metal nanoparticles act as efficient catalysts for the electroless deposition of metals. For example, Ag-nanoparticles catalyze the deposition of silver,[24] Pd-nanoparticles catalyze the deposition of copper or nickel[32] and Au-nanoparticles "seed" the electroless deposition of gold.[33] Recently, the electroless deposition of Ag, using DNA-crosslinked Au-nanoparticle arrays as a catalytic assembly, was employed for the amplified optical detection of DNA.

We have used the catalytic properties of Au-nanoparticles to catalyze the reduction of $AuCl_4^-$ in the presence of hydroxylamine (NH_2OH), to develop amplified microgravimetric detection schemes for the analysis of DNA.[34] Scheme 3(A) shows the method employed to amplify the detection of the target nucleic acid (**7**). The Au-quartz crystal modified with the primer (**6**) is interacted with the analyte sample, (**7**), to yield the double-stranded complex. Reaction of the interface with the biotinylated nucleic acid (**9**), followed by the association of Au-nanoparticle-labeled avidin provides the first step of amplification. The catalytic nanoparticle is linked to the electronic transducer only if the analyte (**7**) is present on the interface.

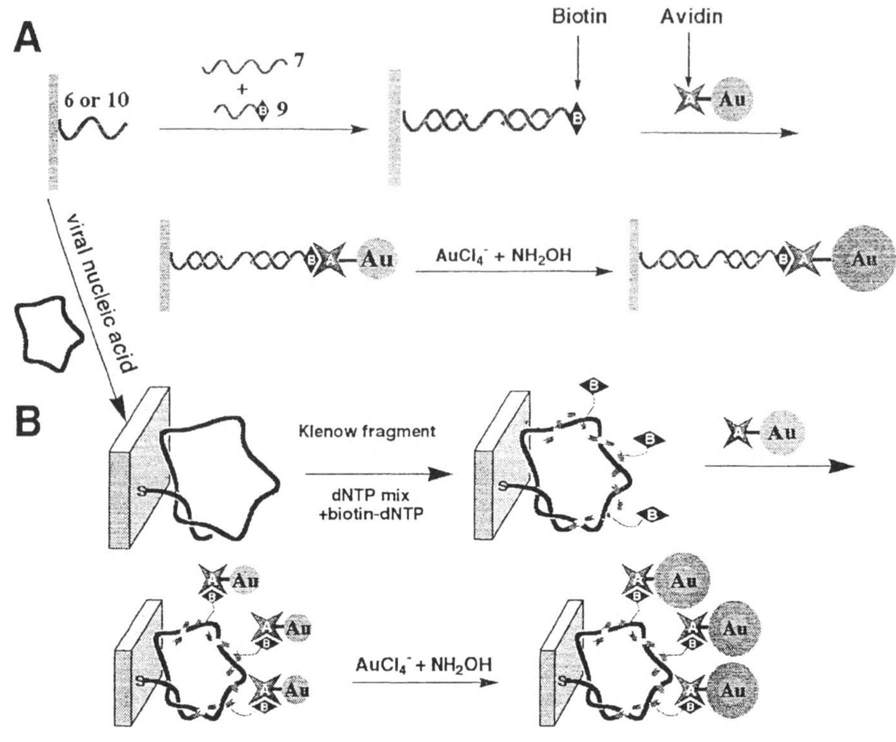

(7) 5'-AGCGTAGGATAGATATACGGTTCGCGC-3'
(7a) 5'-AGCGCTCCAGTGATATACGGTTCGCGC-3'
(6) 5'-TCTATCCTACGCT-$(CH_2)_6$-SH-3'
(9) 5'-**biotin**-GCGCGAACCGTATA-3'
(10) '5-HS- $(CH_2)_6$-CCCCCACGTTGTAAAACGACGGCCAGT-3'
(11) 5' - CTT TTC TTT TCT TTT GGA TCC GCA AGG CCA GTA ATC AAA CG - 3'
(11a) 5' - CTT TTC TTT TCT TTT AGA TCC GCA AGG CCA GTA ATC AAA CG - 3'
(12) 5' - HS-$(CH_2)_6$- CGT TTG ATT ACT GGC CTT GCG GAT C - 3'

Scheme 3. The sensing of DNA by association of Au-nanoparticles and the
subsequent deposition of gold. (A) Detection via a DNA-biotin conjugate. (B)
Detection via replication utilizing a biotin-dNTP conjugate.

Figure 9, curves (a) and (b), shows the frequency changes of the
functionalized Au-quartz crystal upon the sensing of the target nucleic acid, (7),
at concentrations that correspond to 1×10^{-9} M, and 1×10^{-13} M, respectively. The
association of the Au-labeled avidin to the system that analyzes (7), 1×10^{-9} M,

results in a frequency-change of $\Delta f = -87$ Hz, whereas the association of Au-labeled avidin to the interface that analyzes the low concentration of (**7**), 1×10^{-13} M, is invisible and within the noise level of the instrument, ±2 Hz. Addition of the $AuCl_4^-$ salt solution, 5×10^{-4} M, and hydroxylamine, 5×10^{-4} M, results in sharp frequency changes as a result of the electroless deposition of gold on the Au-nanoparticles. Upon the analysis of (**7**) at concentrations of 1×10^{-9} M and 1×10^{-13} M, the crystal frequencies decrease by 191 Hz and 865 Hz, respectively, demonstrating the active amplification of the analysis of (**7**) by the catalytic metal deposition process. Figure 9, curve (c), shows the time-dependent changes in the crystal frequencies upon the analysis of a nucleic acid mutant (**7a**) that includes six-base mismatches as compared to (**7**). It is clear that this nucleic acid is not recognized by the (**6**)-functionalized Au-quartz crystal, and thus the Au-labeled avidin does not bind to the modified-crystal interface. As a result, the secondary catalytic deposition of Au does not occur. The minute frequency change, $\Delta f = -9$ Hz, observed upon the attempt to stimulate the amplified deposition of gold, is attributed to trace amounts of non-specific Au-labeled avidin adsorbate that yields, eventually, a gold precipitate. This frequency change is considered as the noise level of the technique.

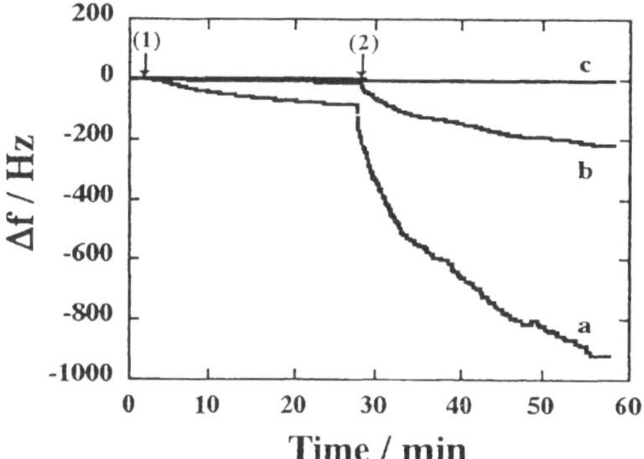

Figure 9. Frequency changes of a 6-functionalized Au-quartz crystal upon the analysis of (a) 7 at 10^{-9} M, (b) 7 at 10^{-13} M, and (c) 7a at 5×10^{-7} M. The Au-avidin conjugate was added at (1), and gold precepitation was induced at (2).

This method was further developed to analyze the complete single-stranded viral gene of M13φ that includes 7229 bases in the c-DNA frame, Scheme 3(B). The primer (**10**) that consists of 27 bases complementary to the M13φ DNA was immobilized onto the Au-quartz crystal to yield the double-stranded assembly. Interaction of the assembly with the phosphorylated base-mixture dNTP that includes biotinylated dCTP, biotin-dCTP at a ratio dCTP:biotin-dCTP of 2:1 (each of the bases in the dNTP mixture at a concentration of 5×10^{-4} M) with polymerase results in the replication of the c-DNA of M13φ. The replication process eventually introduces the biotin-tag on the replicated strand. Association of the Au-nanoparticle-labeled avidin, followed by the interaction of the system with AuCl$_4^-$/NH$_2$OH results in the catalytic deposition of gold on the piezoelectric crystal. Thus, the microgravimetric analysis of M13φ DNA includes three amplification elements: (i) The polymerase-induced replication of the DNA. (ii) The association of the Au-nanoparticle avidin conjugate, and (iii) the biocatalytic electroless deposition of gold on the Au-nanoparticles. Figure 10, curve (a), shows the time-dependent frequency changes of the piezoelectric crystal as a result of the analysis of the M13φ DNA, 1x10^{-15} M, upon the interaction of the double-stranded biotin-labeled replicated assembly with the Au-nanoparticle avidin as a result of the catalytic deposition of gold on the resulting interface. The association of the Au-nanoparticle avidin conjugate results in a frequency change of -11 Hz. The catalytic deposition of gold on the interface reveals a substantial decrease in the crystal frequency, Δf = -165 Hz, revealing the amplifying feature of the analyte process. For comparison, Figure 10, curve (b), shows the results obtained upon reacting the (**10**)-functionalized electrode with calf thymus DNA according to the identical analytical protocol. It is evident that no frequency changes are detected upon interaction of the system obtained after a "replication attempt", with the Au-nanoparticle-avidin conjugate. The catalytic precipitation step of gold in the presence of AuCl$_4^-$ /NH$_2$OH results in a frequency change of ca. -3 Hz, that is attributed to the catalytic deposition of gold on trace amounts of Au-nanoparticle-avidin conjugate that adsorbs non-specifically to the sensing interface. The frequency-change originating from the deposition of gold by non-specific Au-nanoparticle-avidin adsorbates is considered as the noise level for the analytical procedure. Unprecedented sensitivities in the specific detection of DNA by the method were accomplished, and the M13φ DNA at a concentration of 1x10^{-15} M, could be easily detected.

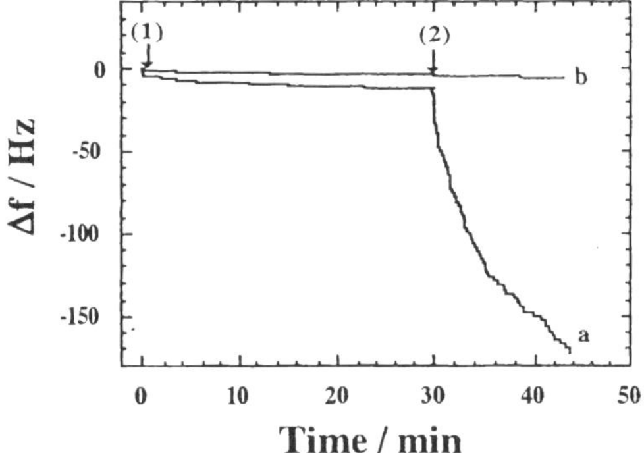

*Figure 10. Frequency changes of a **10**-functionalized Au-quartz crystal upon the analysis of (a) M13φ at 10^{-15} M, (b) calf thymus DNA at 10^{-9} M. The Au-avidin conjugate was added at (1), and gold precepitation was induced at (2).*

Conclusions and Perspectives

Metallic and semiconductor nanoparticles provide unique conductivity, catalytic, optical, photonic and photoelectrochemical properties. This article has addressed means to tailor nanoparticle architectures for sensoric and electronic applications. We, as well as other groups, have demonstrated the ability to apply molecular acceptors and double-stranded DNA as bridging units for these arrays. The nanoscale structure of the assemblies provides high-surface area systems of tunable electronic and optical properties, and the crosslinking of the nanoparticles by molecular receptors leads to new sensor devices. Similarly, the crosslinking of metal and semiconductor nanoparticles by DNA gives novel amplified sensing schemes for DNA, with microgravimetric and optical detection means. The ability to use DNA as a linker opens the way to construct complex assemblies of nanoparticles on surfaces. Furthermore, the DNA units may act as templates for the construction of metallic or insulating wires, or may act as matrices for the incorporation of additional components, e.g. intercalators.

With this basic knowledge, one may envisage the future perspectives of the field. The organization of composite arrays consisting of different metal and semiconductor nanoparticles, or the construction of programmed arrays that include gradients of nanoparticle dimensions, may be envisaged. New functions of such arrays, such as light antennas, may be expected. The functionalization of semiconductor nanoparticles with photoswitchable molecular functionalities could even lead to new optoelectronic memories.

One important future challenge is the organization of addressable nanoparticle architectures. Primary efforts in the photo-patterning of nanoparticle structures have already been reported.[35] The rapid advances in developing scanning microscopy 'writing' techniques[36] suggest that the use of nanoparticle 'inks' is a feasible goal. The progress in the optical addressing of nanometer-sized domains using nearfield scanning optical microscopy (NSOM), indicates that optically-functional nanostructures are viable concepts.

Acknowledgment: These studies are supported by the U.S.-Israel Binational Science Foundation and by the Israel-Japan Research Cooperation Project. The Max-Planck Research Award for International Cooperation (I.W.) is acknowledged. ANS gratefully acknowledges a Valazzi-Pikovsky fellowship.

References

1. *Nanosystems, Molecular Machinery, Manufacturing and Computation*; Drexler, K.E. Ed., Wiley: New York, **1992**.
2. a) Schmid, G.; Chi, L.F. *Adv. Mater.* **1998**, *10*, 515-526; b) Fendler, J.H.; Melfrum, F.C. *Adv. Mater.* **1995**, *7*, 607-632.

3.   a) Shipway, A.N.; Katz, E.; Willner, I. *ChemPhysChem* **2000**, *1*, 18-52; b) Shipway, A.N.; Willner, I. *Chem. Commun.* **2001**, 2035-2045..

4.   a) Tian, Y.; Newton, T.; Kotov, N.; Guldi, D.; Fendler, J.H. *J. Phys. Chem. B*, **1996**, *100*, 8927-8939; b) Hasselbarth, A.; Eychmüller, A.; Eichberger, R.; Giersig, M. Mew, A.; Weller, H. *J. Phys. Chem.* **1993**, *97*, 5333-5340; c) Templeton, A.C.; Wuelfing, M.P.; Murray, R.W. *Acc. Chem. Res.* **2000**, *33*, 27-36.

5.   Link, S.; El-Sayed, M.A. *J. Phys. Chem. B* **1999**, *103*, 8410-8426.

6.   a) Weller, H. *Angew. Chem. Int. Ed.* **1998**, *37*, 1658-1659; b) Weller, H.; Eychmüller, A. *Adv. Photochem.* **1995**, *20*, 165.

7.   a) Brus, L.E. *Appl. Phys. A.* **1991**, *53*, 465-474; b) Alivisatos, A.P. *Science* **1996**, *271*, 933-937.

8.   Mulvaney, P. *Langmuir* **1996**, *12*, 788-800.

9.   Lewis, L.N. *Chem. Rev.* **1993**, *93*, 2693-2730.

10.   Shipway, A.N.; Lahav, M.; Willner, I. *Adv. Mater.* **2000**, *12*, 993-998.

11.   Peng, X.; Wilson, T.E.; Alivisatos, A.P.; Schultz, P.G. *Angew. Chem. Int. Ed. Engl.* **1997**, *36*, 145-147.

12.   Brousseau III, L.C.; Novak, J.P.; Marinakos, S.M.; Feldheim, D.L. *Adv. Mater.* **1999**, *11*, 447-449.

13.   Shipway, A.N.; Lahav, M.; Gabai, R.; Willner, R. *Langmuir* **2000**, *16*, 8789-8795.

14.   a) Marinakos, S.M.; Brousseau III, L.C.; Jones, A.; Feldheim, D.L. *Chem. Mater.* **1998**, *10*, 1214-1219; b) Chung, S.-W.; Markovich, G.; Heath, J.R. *J. Phys. Chem. B* **1999**, *102*, 6685-6687; c) Sano, M.; Kamino, A.; Shinkai, S. *Langmuir* **1999**, *15*, 13-15.

15.   a) Doron, A.; Katz, E.; Willner, I. *Langmuir* **1995**, *11*, 1313-1317; b) Grabar, K.C.; Allison, K.J.; Baker, B.E.; Bright, R.M.; Brown, K.R.; Freeman, R.G.; Fox, A.P.; Keating, C.D.; Musick, M.D.; Natan, M.J. *Langmuir* **1996**, *12*, 2353-2361; c) Sarathy, K.V.; Thomas, P.J.; Kulkarni, G.U.; Rao, C.N.R. *J. Phys. Chem. B* **1999**, *103*, 399-401.

16.   a) Musick, M.D.; Keating, C.D.; Keefe, M.H.; Natan, M.J. *Chem. Mater.* **1997**, *9*, 1499-1501; b) Baum, T.; Bethell, D.; Brust, M.; Schiffrin, D.J. *Langmuir* **1999**, *15*, 866-871; c) Fendler, J.H. *Chem. Mater.* **1996**, *8*, 1616-1624; d) Lvov, Y.M.; Rusling, J.F.; Thomsen, D.T.; Papadimitrakopoulos, F.; Kawakami, T.; Kunitake, T. *Chem. Commun.* **1998**, 1229-1230; e) Shipway, A.N.; Lahav, M.; Willner, I. *Adv. Mater.* **2000**, *12*, 993-998.

17.   Gittins, D.I.; Bethell, D.; Schiffrin, D.J.; Nichols, R.J. *Nature* **2000**, *408*, 67-69.

18.   a) Feldheim, D.L.; Keating, C.D. Chem. Soc. Rev. **1998**, 27, 1-12; b) Sato, T.; Ahmed, H.; Brown, D.; Johnson, B.F.G. *J. Appl. Phys.* **1997**, *82*, 696-701; c) Klein, D.L.; Roth, R.; Kim, A.K.L.; Alivisatos, A.P.; McEuen, P.L. *Nature* **1997**, *389*, 699-701; d) Simon, U.; Schön, G.; Schmid, G. *Angew. Chem. Int. Ed. Engl.* **1993**, *32*, 250-254; e) Andres, R.P.; Bein, T.; Dorogi, M.; Feng, S.; Henderson, J.I.; Kubiak, C.P.; Mahoney, W.; Osifchin, R.G.;

Reifenberger, R. *Science* **1996**, *272*, 1323-1325; f) Sweryda-Krawiec, B.; Cassagneau, T.; Fendler, J.H. *Adv. Mater.* **1999**, *11*, 659-664.

19. Shipway, A.N.; Lahav, M.; Blonder, R.; Willner, I. *Chem. Mater.* **1999**, *11*, 13-15.

20. Asakawa, M.; Ashton, P.R.; Menzer, S.; Raymo, F.M.; Stoddart, J.F.; White, A.J.P.; Williams, D.J. *Chem. Eur. J.* **1996**, *2*, 877.

21. a) Lahav, M.; Shipway, A.N.; Willner, I. *J. Chem. Soc., Perkin Trans. 2* **1999**, 1925-1931; b) Kharitonov, A.B.; Shipway, A.N.; Willner, I. *Anal. Chem.* **1999**, *71*, 5441-5443; c) Kharitonov, A.B.; Shipway, A.N.; Katz, E.; Willner, I. *Rev. Anal. Chem.* **1999**, *18*, 255-260.

22. Lahav, M.; Shipway, A.N.; Willner, I.; Nielsen, M.B.; Stoddart, J.F. *J. Electroanal. Chem.* **2000**, *482*, 217-221.

23. a) Seeman, N.C. *Acc. Chem. Res.* **1997**, *30*, 357-363; b) Niemeyer, C.M. *Angew. Chem. Int Ed. Engl.* **1997**, *36*, 585-587.

24. Braun, E.; Eichen, Y.; Sivan, U.; Ben-Yoseph, G. *Nature* **1998**, *391*, 775-778.

25. a) Storhoff, J.J.; Mirkin, C.A. *Chem. Rev.* **1999**, *99*, 1849-1862; b) Elghanian, R.; Storhoff, J.J.; Mucic, R.C.; Letsinger, R.L.; Mirkin, C.A. *Science* **1997**, *277*, 1078-1081.

26. a) Patolsky, F.; Ranjit, K.T.; Lichtenstein, A.; Willner, I. *Chem. Commun.* **2000**, 1025-1026; b) Taton, T.A.; Mucic, R.C.; Mirkin, C.A.; Lestinger, R.L. *J. Am. Chem. Soc.* **2000**, *122*, 6305-6306.

27. Willner, I.; Patolsky, F.; Wasserman, J. *Angew. Chem. Int. Ed.* **2001**, *40*, 1861-1864.

28. Lewis, F.D.; Kalgutkar, R.S.; Wu, Y.; Liu, X.; Liu, J.; Hayes, R.T.; Miller, S.E.; Wasielewski, M.R. *J. Am. Chem. Soc.* **2000**, *122*, 12346-12351.

29. Steel, A.B.; Herne, T.M.; Tarlov, M.J. *Anal. Chem.* **1998**, *70*, 4670-4677.

30. a) Patolsky, F.; Katz, E.; Bardea, A.; Willner, I. *Langmuir,* **1999**, *5*, 3703-3706; b) Alfonta, L.; Singh, A.K.; Willner, I. *Anal. Chem.,* 2001, *73*, 91-102.

31. Bardea, A.; Katz, E.; Willner, I. *Electroanalysis,* **2000**, *12*, 1097-1106.

32. Richter, J.; Seidel, R.; Kirsch, R.; Mertig, M.; Pompe, W.; Plaschke, J.; Schackert, H.K. *Adv. Mater.* **2000**, *12*, 507-510.

33. Brown, K.R.; Lyon, L.A.; Fox, A.P.; Reiss, B.D.; Natan, M.J. *Chem. Mater.* **2000**, *12*, 314-323.

34. Weizmann, Y.; Patolsky, F.; Willner, I. *Analyst* **2001**, *126*, 1502-1504.

35. a) Vossmeyer, T.; DeIonno, E.; Heath, J.R. *Angew. Chem. Int. Ed. Engl.* **1997**, *36*, 1080-1083; b) Hidber, P.C.; Helbig, W.; Kim, E.; Whitesides, G.M. *Langmuir* **1996**, *12*, 1375-1380.

36. Piner, R.D.; Zhu, J.; Xu, F.; Hong, S.H.; Mirkin, C.A. *Science* **1999**, *283*, 661-663.

Chapter 9

Laser-Induced Alloying in Metal Nanoparticles: Controlling Spectral Properties with Light

Gregory V. Hartland, Steve Guillaudeu, and Jose H. Hodak

Department of Chemistry and Biochemistry, University of Notre Dame,
251 Nieuwland Science Hall, South Bend, IN 46556–5670

Laser excitation of metal nanoparticles can provide enough energy to melt or even fragment the particles. In this article we describe some recent experiments where controlled laser excitation was used to transform core-shell bimetallic particles into the corresponding alloy. Results for Au-Ag particles in solution and in a thin film are presented. Details are given about the excitation energies needed for alloying and how interdiffusion and alloying occur in nanoparticles. The spectral and dynamical properties of bimetallic particles are also discussed — especially as they pertain to our experiments.

Introduction

Metal nanoparticles are a continual topic of interest to chemists and engineers because of their unusual optical properties and their extensive use as catalysts (*1,2*). The bulk of our knowledge about the optical properties of metal particles (both steady-state and time-resolved) comes from single component particles, usually of noble metals. Much less is known about the spectroscopy of multi-component particles, or particles of transition metals, such as: Pt, Pd and Rh. A quantitative understanding of the spectroscopy of single and multi-component transition metal particles would be useful, since it would allow *in situ* measurements of the state of working catalysts. At present one of the limiting factors for developing this understanding is that it is difficult to make multi-component particles with arbitrary combinations of metals, that have a well defined structure (for example, core-shell compared to alloyed) and are in a controlled environment.

In this article the use of laser induced melting to control the structure of bimetallic particles of Ag and Au, is described (*3*). The particles are first synthesized by depositing a shell of one metal over a core of the other, using either radiation-chemistry (*4-9*) or chemical reduction techniques (*10-12*). Alloyed nanoparticles are subsequently made by heating the core-shell particles with a moderately intense nanosecond or picosecond laser beam. These experiments allow us to investigate how the spectra of the particles change as their structure changes. They also provide insight into the sequence of events that occur after laser excitation of metal particles, and the mechanism and timescales for interdiffusion in nanoscale systems. The advantages of this technique for synthesizing alloyed nanoparticles are also discussed.

Experimental

All the experiments described below were performed with bimetallic particles that were made by depositing a shell of one metal onto a core of a different metal. Two types of core particles are used in our laboratory: 15 nm diameter Au cores and 12.5 nm diameter Pt cores. The Au particles were made by reduction of $HAuCl_4$ by citrate (*13*), and the Pt cores were made by reduction of $PtCl_2(H_2O)_2$ with H_2 (*14*). The shells were subsequently deposited by adding a salt of another metal to a solution of the core particles, and using γ-ray induced reduction to deposit the second metal onto the core (*4-9*). In this scheme, the radicals produced by γ-irradiation do not have enough energy to directly reduce the metal ions in solution. However, they can deposit electrons onto the core particles, and when sufficient charge has built up the metal ions in solution undergo reduction at the particle surface. Au-Ag core-shell particles can also be made by using a chemical reductant (NH_2OH) to deposit the shell (*10,11*). No difference could be observed between the spectra of the Ag-Au core-shell particles produced by γ-irradiation or by chemical reduction (for the same particle composition). Some experiments were performed with Au-Ag core-shell particles that had been coated with silica. The particles used in these experiments were prepared by NH_2OH reduction. Before coating the particle solution was extensively dialyzed remove as much excess NH_2OH as possible. Following dialysis the pH was adjusted to 5 and the silica coupling agent 3-

aminopropyl-trimethoxy silane was added (*15*). After stirring the pH was increased to 11.5 and Na_2SiO_3 was added to form the silica shell. The thickness of the shell was increased to approximately 120 nm by adding ethanol to the solution (*16*).

The laser induced melting experiments were performed with either a nanosecond (Continuum Surelite I, 5 ns pulsewidth) or picosecond (Continuum PY-61, 30 ps pulsewidth) Nd:YAG laser. Both lasers operate at 532 nm with a 10 Hz repetition rate. In a typical experiment 2 mL of an aqueous solution containing 2×10^{-4} M of metal was irradiated for 15 minutes. The solution was gently stirred and the power absorbed by the sample was monitored throughout the experiment. It is important to note that the laser powers needed to achieve a given level of alloying are different for the nanosecond and picosecond lasers (*3*). This will be discussed in more detail below. Time-resolved experiments were performed with a regeneratively amplified Ti:Sapphire laser system, using a pump-probe experimental configuration. The pump laser pulses (390 nm) were derived from the second-harmonic of the Ti:Sapphire fundamental and the probe pulses were taken from a white-light continuum. This laser system has a time-resolution of ca. 200 fs. The spectral changes in the solutions were monitored with a standard UV/vis spectrometer (Perkin-Elmer Lambda 6). Experiments were also performed with thin films. The films were made by dipping a glass coverslip into a 1% w/w solution of a cationic polymer (PDDA) for ca. 5 minutes, and then placing the coated coverslip in a concentrated solution of particles. The coated coverslips were allowed to dry before being exposed to the laser pulses. The changes in the morphology of the films induced by the laser were subsequently analyzed by tapping-mode AFM (Digital Instruments Nanoscope IIIa, Multimode Scanning Probe Microscope).

Theoretical Considerations

The main concerns of this paper are the long time processes of melting and interdiffusion that occur after laser excitation of bimetallic particles. However, in order to understand the differences between the nanosecond and picosecond excitation experiments described below, it is important to understand the photophysics of metal particles. Figure 1 shows the spectra of different sized Au particles. The strong peak at ca. 520 nm is the plasmon band — which is a collective oscillation of the conduction band electrons across the particle surface (*1*). The plasmon band is unique to small metal particles, and is responsible for the brilliant colors of solutions of Au, Ag or Cu particles. The broad feature to the blue of the plasmon band arises from the *5d → 6sp* interband transitions of Au (*17,18*). The broadening of the plasmon band for the small particles is due to electron scattering at the particle surface (*19-21*). The red-shift and broadening for the large 120 nm particles is due to an increased contribution from scattering to the extinction coefficient of the particles (*22*).

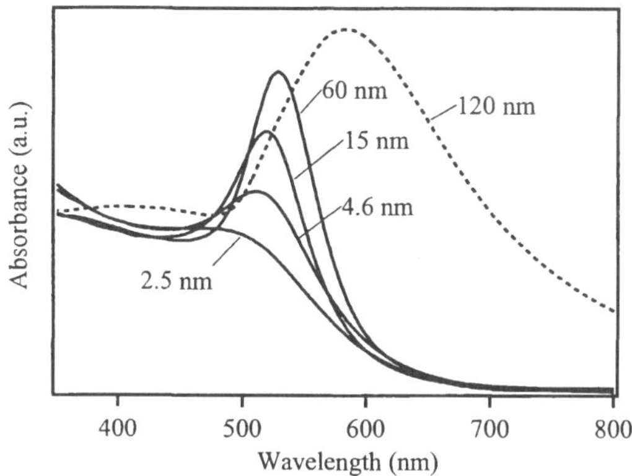

Figure 1. Extinction spectra of different sized Au particles in aqueous solution. The [Au] is approximately the same for each spectrum.

Our experiments are performed with either 532 nm or 390 nm laser pulses, which are resonant to the plasmon oscillation or the interband transitions, respectively. The energy deposited into the electrons is rapidly (~200 fs) spread over the entire electron distribution, creating a hot electron distribution that can have a temperature of several thousand K (*23-25*). The hot electrons subsequently equilibrate with the phonon modes of the particle by electron-phonon scattering, creating a hot electron-phonon system.

The key parameter that determines the temperature rise of the lattice is the amount of energy adsorbed per laser pulse per atom. Assuming that there is no energy loss from the electrons during light absorption, the initial electronic temperature (T_e^i) created by the laser is:

$$\left(T_e^i\right)^2 = T_o^2 + \frac{2nhcN_A}{\lambda\gamma} \,, \tag{1}$$

where h, c and N_A have their usual meanings, n is the number of photons absorbed per atom, λ is the laser wavelength, T_o is the ambient temperature, and γ is related to the electronic heat capacity by $C_e(T_e) = \gamma T_e$, where T_e is the electronic temperature (*26*).

110

The energy absorbed by the electrons is subsequently equilibrated between the electrons and phonons. Assuming that there is no energy loss to the solution during this process, the final temperature of the system T_f is given by:

$$\frac{1}{2}\gamma\left(T_f^{\,2}-(T_e^i)^2\right)=-C_l\left(T_f-T_o\right) \qquad (2)$$

where C_l is the heat capacity of the lattice.

Figure 2 shows a plot of T_f and T_e^i versus n for 532 nm laser pulses. The horizontal line indicates the melting point of bulk Au. These results show that the final temperature approaches the melting point of Au when $n \approx 0.1$, i.e., for a 15 nm diameter Au particle absorption of 10^4 photons with λ = 532 nm supplies sufficient energy to melt the particle. The actual laser power needed to do this depends on the absorption cross-section of the particles at 532 nm, and the time scale for energy dissipation to the solvent compared to the laser pulsewidth.

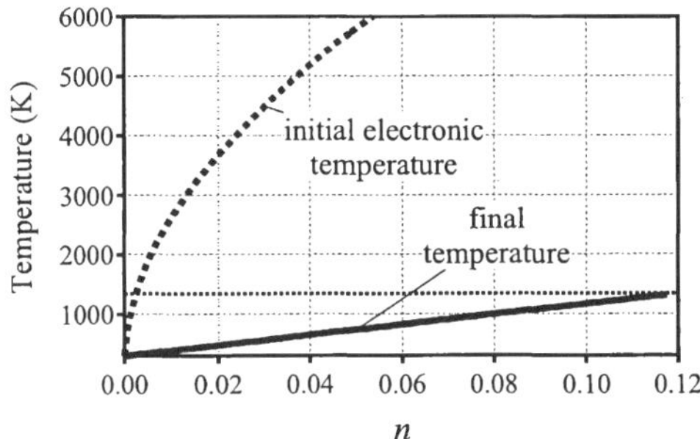

Figure 2. Calculated values of T_e^i and T_f versus the number of photons absorbed per atom (n) for 532 nm laser pulses.

The time-scale for heat dissipation from the particles is very important in determining the overall temperature rise of the lattice. A simple way to model this process is to use Newton's Law for cooling, where the rate of energy exchange between the electrons and the phonons, and the phonons and the environment is assumed to only depend on their temperature differences (*27*). Specifically, if T_e is the electronic temperature and T_l is the lattice temperature, then (*27-30*):

$$C_e(T_e)\frac{\partial T_e}{\partial t} = -g(T_e - T_l) + F(t) \tag{3a}$$

$$C_l \frac{\partial T_l}{\partial t} = g(T_e - T_l) - \frac{T_l - 298}{\tau}. \tag{3b}$$

In these equations $F(t)$ is a source term (i.e, the laser pulse), g is the electron-phonon (*e-ph*) coupling constant, τ gives the time-scale for energy exchange between the particle and its environment, and we have assumed that the temperature of the environment is constant at 298 K. The *e-ph* coupling coupling constant has been measured for a wide variety of single component metal nanoparticles (*23-25*), however, there are very few measurements for bi-metallic particles (which are the major concern of this paper). We have recently performed time-resolved experiments for Pt-Au particles that show that their *e-ph* coupling constant is simply an average of the *e-ph* coupling constants for pure Pt and Au, weighted by the fraction of electronic states for each (*31*). Thus, for the calculations described below we will use an average of the *e-ph* coupling constants for Ag and Au.

The time-scale for energy transfer from the particles to the solvent can be probed by transient absorption experiments. Figure 3 shows the results of experiments for 15 nm and 60 nm diameter Au particles that were performed with equal energy 390 nm pump laser pulses.

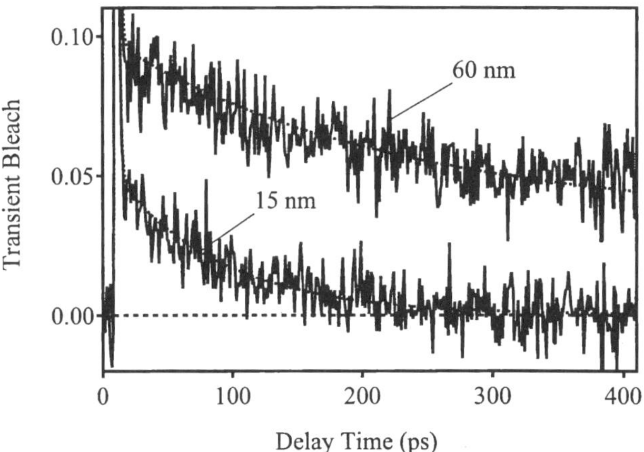

Figure 3. Time resolved experiments with 15 nm and 60 nm diameter Au particles. The observed decays correspond to energy transfer to the solvent.

At 390 nm the absorbance per Au atom is the same for the 15 nm and 60 nm diameter particles, thus, the increase in lattice temperature is approximately the same for both samples. The key point to note from these experiments is that the time-scales for energy transfer are different for the 15 nm and 60 nm particles. Energy transfer is faster for the 15 nm particles because they have a greater surface-to-volume ratio (heat transfer occurs through the surface of the particles). We have also noticed that the exact value of the energy relaxation time depends on the pump laser power. However, in general heat transfer from the particles to the solution occurs on a time-scale of hundreds of picoseconds.

Figure 4 shows simulations for the evolution of the electronic and lattice temperatures for 30 ps and 5 ns laser pulses. The parameters used where $C_l=$ 25.42 J K^{-1} mol^{-1}, $\gamma = 6.74 \times 10^{-4}$ J K^{-2} mol^{-1}, $g = 3.1 \times 10^{11}$ W K^{-1} mol^{-1} and $\tau =$ 400 ps (estimated from Figure 3). The magnitude of $F(t)$ was scaled to correspond to $n = 0.1$. Note that T_e and T_l have the same time dependence for ns excitation (they rise and fall with the laser pulse), whereas, for ps excitation T_e and T_l evolve very differently while the pulse is on. This is because the pulsewidth in the ps experiments is comparable to the time scale for electron-phonon coupling (a few ps in these simulations), so that the electrons and phonons are not in thermal equilibrium during the laser pulse.

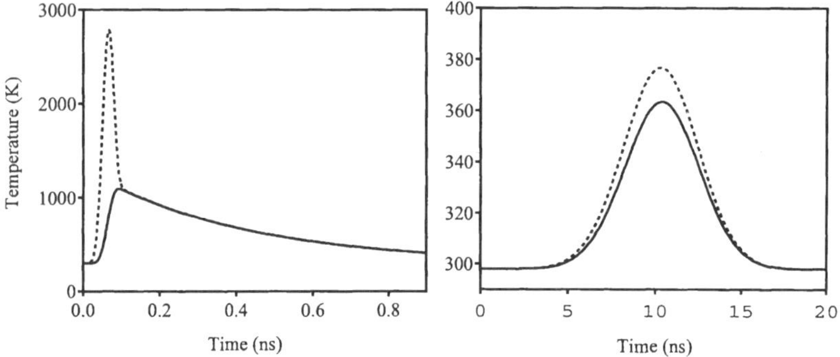

Figure 4. Simulation of the electronic and lattice temperatures for picosecond (left) and nanosecond (right) laser pulses.

It is also important to note that the lattice temperatures that can be achieved with ns excitation are much lower than those for an equal energy ps pulse (i.e., ps pulses are more efficient for heating). This is because excitation and energy dissipation to the environment compete for the ns pulses. Also, the initial value of the lattice temperature in the ps experiments is very close to the value calculated by equations (1) and (2) (i.e., without considering energy dissipation), whereas, the lattice temperature in the ns experiments depends on the time scale for energy transfer to the solvent. Figure 5 shows a plot of how the maximum lattice temperature varies with the energy transfer time for the nanosecond simulations described above (i.e., 5 ns pulses with $n = 0.1$).

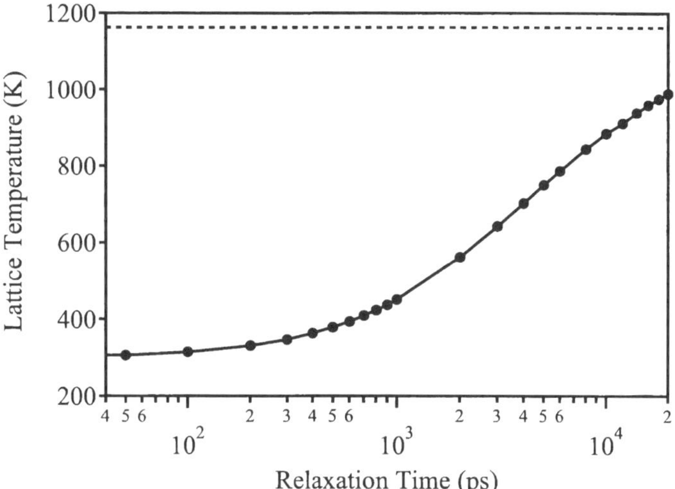

Figure 5. Maximum calculated lattice temperatures for a 532 nm nanosecond laser pulse (n = 0.1) versus the time scale for energy transfer to the solvent. The dashed line represents the lattice temperature in the limit of no relaxation.

For the core-shell nanoparticles, the increase in lattice temperature caused by laser excitation will increase the rate of interdiffusion of the two metals. If the particles remain hot enough for long enough, they will eventually be transformed into an alloy.

114

Experimental Results

Particles in Aqueous Solution

Figure 6 shows spectra for Au-Ag core-shell nanoparticles in aqueous solution (1:0.5 ratio of Au:Ag) that have been excited with different intensity 532 nm, 30 ps laser pulses. The solutions were irradiated for 15 minutes. The adsorbed energies per pulse were E_{ads} = 0.1 mJ/pulse for spectra (b) and 1 mJ/pulse for spectra (c). Under our experimental conditions, these energies correspond to n = 0.016 and n = 0.092 photons adsorbed per pulse per atom, respectively. The initial spectrum shows two plasmon bands, which is characteristic of a core-shell structure (*1*). The single peak in spectrum (c), which is in between the plasmon bands of Ag (390 nm) and Au (520 nm) is characteristic of an alloy (*32*). Note that further irradiation, or irradiation at higher power, did not change the peak position for this sample. This implies that spectrum (c) corresponds to a fully alloyed sample. At very high laser power (E_{ads} > 4 mJ/pulse for 30 ps pulses) the peak intensity decreases, which is consistent with fragmentation of the particles. This has been confirmed TEM measurements, which show that small (< 10 nm) particles are produced at high powers (*3*). At present we do not know the composition of the fragments, i.e., when you blow apart Au-Ag core-shell particles, are the fragments produced predominantly Ag or a mixture of Au and Ag?

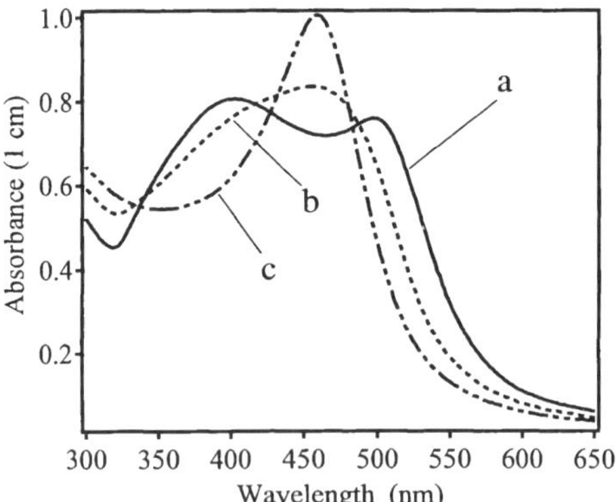

Figure 6: Absorption spectra of Au-Ag core-shell particles before (a) and after irradiation with ps laser pulses: (b) n = 0.016; (c) n = 0.092.

The observed plasmon band maximum of 460 nm for spectrum (c) does not quantitatively agree with the results of Link and co-workers (*32*). In their experiments alloyed nanoparticles were prepared by simultaneous reduction of Ag and Au salts. It is not clear whether the difference in the peak positions is due to incomplete alloying in our samples or those of Link and co-workers. Note that Treguer et al. have used γ-irradiation to simultaneously reduce Ag and Au salts (*33*). They found that alloyed particles are prepared at high dose, but core-shell particles are prepared at low dose. These differences arise from a competition between particle growth and electron transfer from zero-valent Ag (produced by irradiation) to Au(III) ions in solution.

Spectrum (b) in Figure 6 ($n = 0.016$ photons adsorbed per pulse per atom) corresponds to particles that are in between a core-shell and an alloyed structure. Further irradiation at the laser power used in this experiment does not produce a significant change in the spectrum. However, we are unwilling to state that the particles will never become completely alloyed, if we were to continue to irradiate for long enough! On the other hand, the changes in Figure 6 spectrum (b) show that there is significant interdiffusion of Ag and Au at this power. Note that the results in Figure 2 show that the maximum lattice temperature for $n = 0.016$ is much lower than the bulk melting points of Ag or Au.

The differences between spectra (b) and (c) in Figure 6 arise because interdiffusion in metals is an activated process (*34*) — higher laser powers produce higher temperatures and, therefore, more extensive alloying for a given irradiation time. It is important to note that the alloying process takes many laser pulses to complete. This can also be understood from Figure 4: because of coupling to the environment, the particles do not stay hot for long enough for complete mixing of the two metals. Each laser pulse heats up the particles, allowing partial mixing to occur. As the particles cool, the rate of interdiffusion decreases and eventually effectively stops. For a 15 minute irradiation time we find that the threshold power for creating "fully" alloyed nanoparticles is $n = 0.092$ for 30 ps excitation pulses, and $n = 0.35$ for 5 ns laser pulses. The higher threshold for the ns excitation experiments is due to competition between excitation and energy dissipation to the solvent. Specifically, for equal adsorbed energies, ns pulses yield a lower overall increase in the lattice temperature (see Figure 4) and therefore, less mixing between Ag and Au.

Particles in Thin Films

Similar experiments were performed with Au-Ag core-shell nanoparticles deposited onto a glass substrate. Two types of particles were used in these experiments: "bare" particles that are stabilized in solution by adsorbed ions (such as citrate), and silica coated particles - prepared using the recipe described in Refs. (*15,16*) and the experimental section of this paper. Silica coating metal particles is extremely popular, primarily because it stabilizes them in solution (high volume fractions of silica coated metal particles can be obtained without inducing irreversible flocculation). Figure 7 shows UV/visible absorption spectra of thin films of coated and uncoated Au-Ag core-shell particles, both before and after irradiation with 30 ps laser pulses.

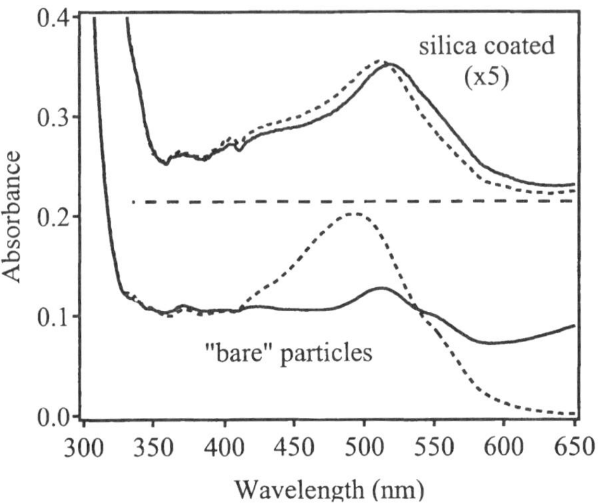

Figure 7: Absorbance spectra of Au-Ag nanoparticle films. The solid lines are before laser induced melting, and the dashed lines are after excitation. The spectra have been off-set for clarity. Note the rising absorbance at 650 nm for the unirradiated "bare" particles.

There are several points to note. First, before irradiation, the bare particles (no silica coating) have significant absorption throughout the visible region of the spectrum, into the near-IR. This extended plasmon band arises from interactions between particles, and is characteristic of thin films formed from close packed metal particles (*35*). On the other hand, the thin films of silica coated particles do not show this extended plasmon band, presumably because the ~120 nm silica shell separates the particles on the film (*36*). Note that the absorbance of the silica coated particles is weaker, simply because there is less metal in the film. Second, after irradiation, both the silica coated and uncoated particles display a single peak in the UV/vis absorption spectrum, close to where the plasmon band of the alloyed particles occurs in solution. (We do not expect exactly the same peak positions for the films and solutions due to the effect of the substrate.) Surprisingly, the extended plasmon band due to interactions between the particles in the "bare" sample has disappeared after irradiation.

A picture of the films is shown in Figure 8. The purplish film on the right consists of the "bare" Au-Ag core-shell particles and the red film on the left contains the silica coated particles. The orange spot in the middle of each film is the area irradiated by the laser. Note that the color of the irradiated film is similar for the "bare" and silica coated particles. The changes induced by laser excitation were also analyzed by AFM. Figure 9 shows line-scan images obtained by tapping mode AFM of the uncoated Au-Ag particle films before and after irradiation. The images are presented as line-scans to emphasize the size differences caused by laser excitation. These experiments show that the uncoated sample has been transformed from closely packed particles with an average diameter of 20 nm (as determined by height analysis) to separated particles with an average diameter of 30 nm. This change in height corresponds to a factor of approximately three increase in the size of the particles.

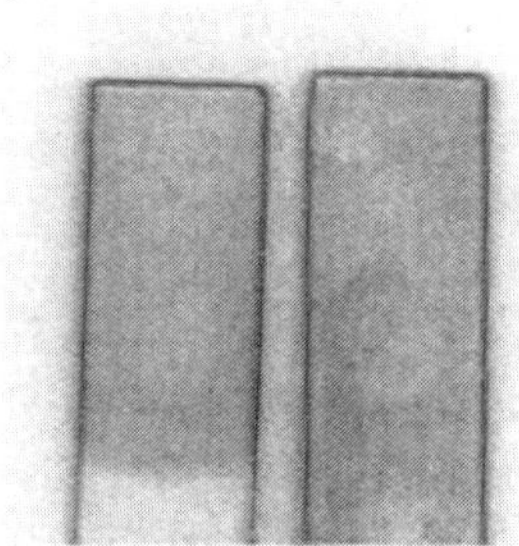

Figure 8: Picture of the silica coated and bare Au-Ag core-shell nanoparticles films after irradiation by 532 nm laser.

AFM analysis of the silica coated particles shows that laser irradiation does not cause any change in the size of these particles. We interpret these results to show that during laser irradiation of the uncoated particles, neighboring particles that are initially touching coalesce together. This fusion process creates larger particles and also increases the average distance between particles. This is shown schematically in Figure 10. The increase in the average distance between particles reduces their interactions and, so, decreases the intensity of the extended plasmon band, which arises from particle interactions (*35,36*). Similar effects can be observed in films of single component metal particles, see for example, Ref, (*37*). We have also done these experiments with pure Au films, and find that the color of the film changes from purple (characteristic of interacting metal particles) to the ruby red color of isolated Au particles. At

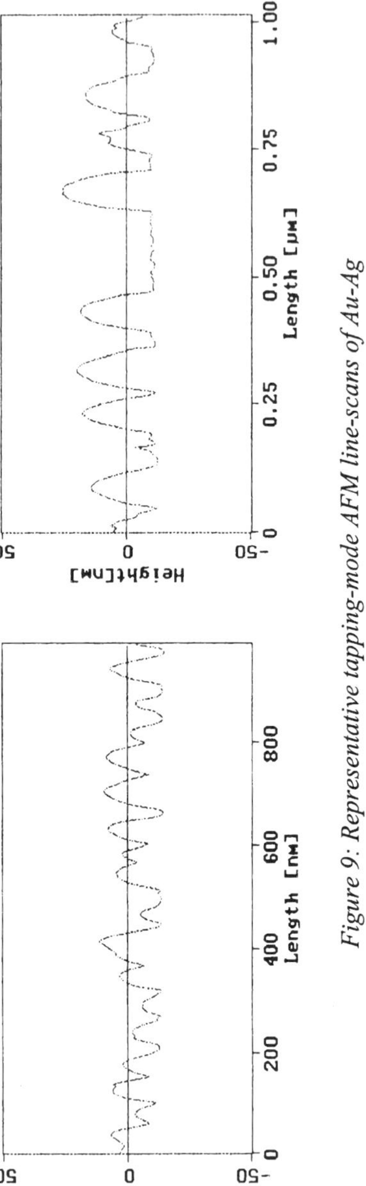

Figure 9: Representative tapping-mode AFM line-scans of Au-Ag nanoparticle films before (left) and after (right) laser irradiation.

present there are several outstanding questions concerning laser-induced melting and coalescence in films: (1) What are the limits of particle growth? Does continued laser excitation lead to increased particle sizes, or is there a maximum

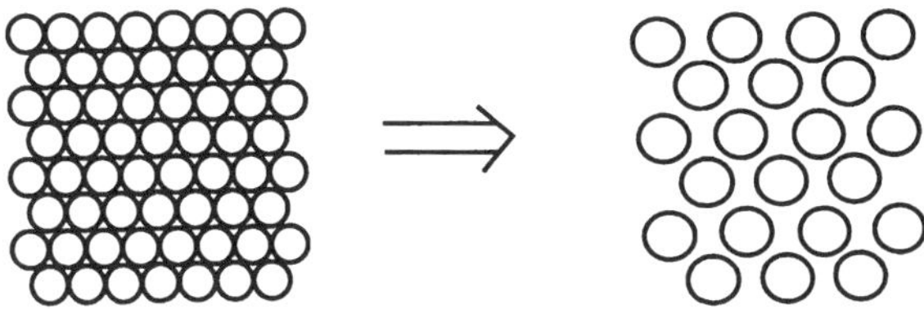

Figure 10: Schematic of laser induced coalescence in metal nanoparticle films. The increase in average size translates into an increase in the average distance between particles in the 2-D film.

size that is dictated by the initial spacings of the particles? (2) How do the properties of the substrate (such as its thermal conductivity and surface chemistry) affect particle coalescence. (3) Can we create non-spherical shapes on surfaces, such as wires? A way that one can imagine that this could happen is by laser annealing particles that are aligned, for example, along a dislocation on the surface. Experiments are currently being carried out to address these issues.

Summary and Conclusions

Laser studies of metal particles have traditionally focussed on measuring fundamental properties of these materials, such as their *e-ph* coupling constant (*23-25,38*). Recently it has been realized that moderately intense (milli-joule) laser pulses can supply enough energy to melt metal particles. This can be used to change the shape of the particles (*39-41*) (i.e., from rods to spheres), as well as their internal structure (alloys can be created from core-shell particles) (*3*). The advantage of this technique for processing particles is that the heating is very selective: the particles can be melted without significantly perturbing the environment. Thus, metal particles in aqueous solution remain in aqueous solution after heating, and particles in thin films can be alloyed without affecting the film.

In this paper we have outlined some of the basic photophysics of metal particles pertaining to laser induced heating and alloying. Specifically, the level of excitation needed to melt the particles, and the differences between nanosecond and picosecond laser pulses were investigated. The main conclusions from this analysis are that for 532 nm excitation, absorption of 0.1 photons per atom (for Au particles) provides enough energy to melt the

120

particles. The temperatures reached for picosecond excitation are higher than those for nanosecond excitation because of competition with energy transfer to the environment for the nanosecond pulses.

We believe that the most exciting prospect for laser-induced melting is to use this technique to create new materials. Specifically, it has been shown that core-shell particles can be made with Pt, Ag or Au as the core, and shells of Pt, Pd, Ag, Au, Cd, Hg or Pb using radiation chemistry (*4-9*). These particles can then be transformed into a homogeneous alloy by laser-induced melting. This gives access to a much wider range of compositions — with better structural control — than conventional co-reduction synthetic techniques (*32,33*). The alloyed particles that can be made by this method will have unique optical properties that differ from the optical properties of core-shell or single component metal particles. They will also have unique catalytic properties: bi-metallic catalysts often have different reactivities and selectivities than single component catalysts (*42,43*). The ability to control the color of the material with light means that laser-induced melting may be useful in application such as optical data storage, and encoding information on materials (such as bank notes or credit cards) for security. Other uses of alloyed particles created by laser-induced melting include synthesis of materials that can be used in magnetic recording (*44*) and in optical coatings (for example for gratings).

Acknowledgements: The work described in this paper was supported by the National Science Foundation by Grant No. CHE98-16164 (GVH and JHH), and by an REU Grant to the Department of Chemistry and Biochemistry. We would like to thank Dr. T. Dhanasekaran for his help with the AFM measurements, and Prof. Dani Meisel of the Notre dame Radiation laboratory for allowing us to use this facility.

References

1. Kreibig, U.; Vollmer, M. *Optical Properties of Metal Clusters*; Springer: Berlin, 1995.
2. *Clusters and Colloids: From Theory to Application*; Schmid, G., Ed.; VCH: Weinheim, 1994.
3. Hodak, J. H.; Henglein, A.; Giersig, M.; Hartland, G. V. *J. Phys. Chem. B,* **2000**, *104*, 11708.
4. Mulvaney, P.; Giersig, M.; Henglein, A. *J. Phys. Chem.* **1993**, *97*, 7061.
5. Henglein, A.; Brancewicz, C. *Chem. Mater.* **1997**, *9*, 2164.
6. Katsikas, L.; Gutierrez, M.; Henglein, A. *J. Phys. Chem.* **1996**, *100*, 11203.
7. Henglein, A. *J. Phys. Chem. B* **2000**, *104*, 2201.
8. Henglein, A.; Giersig, M. *J. Phys. Chem. B* **2000**, *104*, 5056.
9. Henglein, A. *J. Phys. Chem. B* **2000**, *104*, 6683.
10. Morriss, R. H.; Collins, L. F. *J. Chem. Phys.* **1964**, *41*, 3357.

11. Bright, R. M.; Walter, D. G.; Musick, M. D.; Jackson, M. A.; Allison, K. J.; Natan, M. J. *Langmuir* **1996**, *12*, 810.

12. Hostetler, M. J.; Zhong, C.-J.; Yen, B. K. H.; Anderegg, J.; Gross, S. M.; Evans, N. D.; Porter, M.; Murray, R. W. *J. Am. Chem. Soc.* **1998**, *120*, 9396.

13. Enüstün, B. V.; Turkevich, J. *J. Am. Chem. Soc.* **1963**, *85*, 3317.

14. Ahmadi, T. S.; Wang, Z. L.; Green, T. C.; Henglein, A.; El-Sayed, M. A. *Science* **1996**, *272*, 1924.

15. Liz-Marzan, L. M.; Giersig, M.; Mulvaney, P. *Langmuir* **1996**, *12*, 4329.

16. Ung, T.; Liz-Marzan, L. M.; Mulvaney, P. *Langmuir* **1998**, *14*, 3740.

17. Ehrenreich, H.; Philipp, H. R.; *Phys. Rev.* **1962**, *128*, 1622.

18. Johnson, P. B.; Christy, R. W. *Phys. Rev. B* **1972**, *6*, 4370.

19. Doremus, R. H. *J. Chem. Phys.* **1965**, *42*, 414.

20. Kraus, W. A.; Schatz, G. C. *J. Chem. Phys.* **1983**, *79*, 6130.

21. Kreibig, U.; Genzel, U. *Surf. Sci.* **1985**, *156*, 678.

22. Van de Hulst, H. C. *Light Scattering by Small Particles*, Dover: New York, 1981.

23. Link S.; El-Sayed M. A. *J. Phys. Chem. B* **1999**, *103*, 8410.

24. Hodak, J. H.; Henglein, A.; Hartland, G. V. *J. Phys. Chem. B,* **2000**, *104*, 9954.

25. Voisin, C.; Del Fatti, N.; Christofilos, D.; Vallée, F. *J. Phys. Chem. B,* **2001**, *105* 2264.

26. Ashcroft, N. W.; Mermin, N. D. *Solid State Physics,* Harcourt Brace: Orlando, 1976.

27. Kakac, S.; Yenev, Y. *Heat Conduction*; Taylor and Francis: Washington, 1993.

28. Belotskii, E. D.; Tomchuk, P. M. *Surf. Sci.* **1990**, *239*, 143.

29. Belotskii, E. D.; Tomchuck, P. M. *Int. J. Electron.* **1992**, *73*, 955.

30. Sun, C.-K.; Vallée, F.; Acioli, L. H.; Ippen, E. P.; Fujimoto, J. G. *Phys. Rev. B* **1993**, *48*, 12365.

31. Hodak, J. H.; Henglein, A.; Hartland, G. V. *J. Chem. Phys.* **2001**, *114*, 2760.

32. Link S.; Wang, Z. L.; El-Sayed, M. A. *J. Phys. Chem. B* **1999**, *103*, 3529.

33. Treguer, M.; de Cointet, C.; Remita, H.; Khatouri, J.; Mostafavi, M.; Amblard, J.; Belloni, J.; de Keyzer, R. *J. Phys. Chem. B* **1998**, *102*, 4310.

34. Tu, K. N.; Mayer, J. W.; Feldman, L. C. *Electronic Thin Film Science for Electrical Engineers and Materials Scientists*; Macmillan: New York, 1992.

35. Shipway, A. N.; Katz, E.; Willner, I. *Chem. Phys. Chem.* **2000**, *1*, 18.

36. Ung, T.; Liz-Marzan, L.; Mulvaney, P. *J. Phys. Chem. B* **2001**, *105*, 3441.

37. Safonov, V. P.; Shalaev, V. M.; Markel, V. A.; Danilova, Y. E.; Lepeshkin, N. N.; Kim, W.; Rautian, S. G.; Armstrong, R. L. *Phys. Rev. Lett.* **1998**, *80*, 1102.

38. Zhang, J. Z. *Acc. Chem. Res.* **1997**, *30*, 423.

39. Takami, A.; Kurita, H.; Koda, S. *J. Phys. Chem. B* **1999**, *103*, 1226.

40. Chang, S. S.; Shih, C. W.; Chen, C. D.; Lai, W. C.; Wang, C. R. C. *Langmuir* **1999,** *15,* 701.

41. Link, S.; Burda, C.; Mohamed, M. B.; Nikoobakht, B.; El-Sayed, M. A. *J. Phys. Chem. A* **1999,** *103,* 1165.

42. Nashner, M. S.; Frenkel, A. I.; Adler, D. L.; Shapley, J. R.; Nuzzo, R. G. *J. Am. Chem. Soc.* 1997, *119,* 7760.

43. Schmid, G.; West, H.; Mehles, H.; Lehnert, A. *Inorg. Chem.* **1997,** *36,* 891.

44. Black, C. T.; Murray, C. B.; Sandstrom, R. L.; Sun, S. H. *Science* **2000,** *290,* 1131.

Chapter 10

A New Method for Manufacturing Dye-Sensitized Solar Cells on Plastic Substrates

Henrik Lindström, Gerrit Boschloo, Sten-Eric Lindquist, and Anders Hagfeldt[*]

Department of Physical Chemistry, Uppsala University, Box 532, 75121 Uppsala, Sweden

A new method for the preparation of nanostructured TiO_2 films, based on the compression of a TiO_2 powder is presented, and its relevance for the production of dye-sensitized solar cells is discussed. By applying a pressure (>200 kg cm^{-2}) on a film of TiO_2 powder, a mechanically stable nanostructured film is formed, which is suitable as substrate for dye-sensitized solar cells without the necessity of a high temperature sintering step. Using Degussa P25 TiO_2 powder, the so called N719 dye, I^-/I_3^- in 3-methoxypropionitrile as redox electrolyte, and thermally platinized conducting glass as counterelectrode, we have achieved an overall efficiency of $4.5 \pm 0.3\%$ for solar cells using conducting glass and $4.0 \pm 0.5\%$ for cells using conducting plastic. In both cases no high temperature step was applied. The new method allows for a high throughput, continuous roll-to-roll production process, where a roller mill is used for the compression step. Currently, interconnected modules and monolithic cells prepared with the press technique are being studied.

Introduction

In the near future we can expect an increasing number of electronic devices in which molecules replace solid state materials. The dye-sensitized solar cell (DSC) is an example of such a device: dye molecules replace inorganic semiconductor materials as the solar light absorber. In the most studied DSC system, ruthenium(II) dyes are adsorbed at a nanostructured TiO_2 substrate and in contact with a liquid electrolyte containing an I^-/I_3^- redox couple (*1-3*). The conversion of light into electric energy in a DSC works as follows: (1) The dye absorbs photons, at a rate of about 1 s^{-1} in full sunlight. (2) The excited dye injects an electron into the conduction band of TiO_2, a process that occurs on the femtosecond timescale. (3) The oxidized dye molecule is reduced by iodide (ns-μs). (4) Electrons in the nanostructured TiO_2 are transported to the back contact and extracted as a current. (5) Triiodide is reduced to iodide at the counterelectrode.

The results obtained with DSCs are encouraging: after a relatively short development period of about 10 years, certified solar cell efficiencies exceeding 10% have been obtained (*3,4*). In order to compete with solid state solar cells, however, DSCs need to be either more efficient, or much cheaper. (In both cases the long-term stability also needs to be improved to be comparable to that of solid-state devices.) Currently DSCs are manufactured by a method where separate glass sheets are processed through ten to fifteen steps before a complete solar cell is obtained. The production steps include screen-printing and multiple firing steps to temperatures of up to 450°C (*5*). In an effort to make manufacturing much faster and cheaper, we developed a deposition technique for thin films where the main process step is the application of a high pressure. This technique has many advantages over existing techniques, the most important ones being the possibility of very high processing speed and the fact that the firing step is no longer needed.

In figure 1 a schematic view is shown of the new manufacturing technique for nanostructured films. The top part shows the approach using a static press. A powder, suspended in a volatile solvent is spread on a substrate. After evaporation of the solvent, the powder film is pressed. The result is an adherent, mechanically stable porous film. The bottom part shows the continuous method. After application of a powder suspension onto a roll conducting plastic and evaporation of the solvent, a roller mill is used to press the film during a very short time.

Experimental

TiO₂ Film Preparation

TiO_2 powder (Degussa P25) was added to ethanol to a concentration of 20 wt%, followed by stirring. The resulting suspension was applied onto a substrate

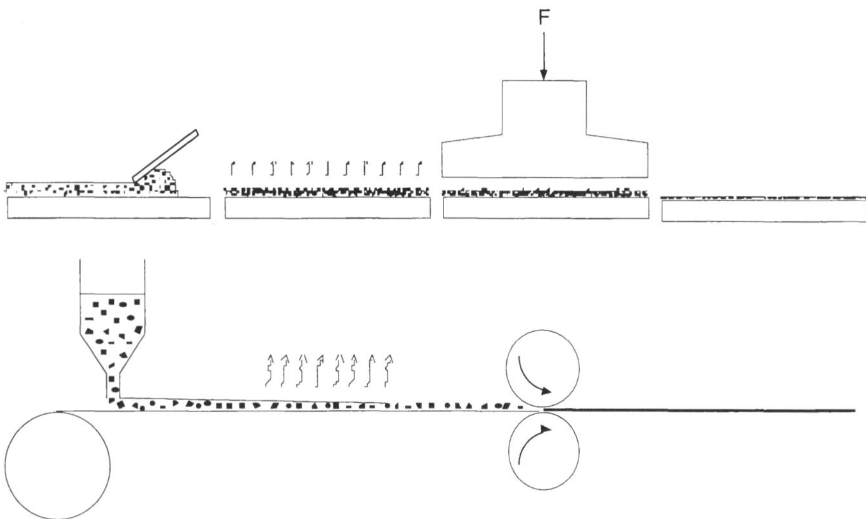

Figure 1. Schematic view of the manufacturing of nanostructured films by compression. Above: static press, below continuous pressing using a roller mill.

by doctor blading using scotch tape as frame and spacer. Two types of substrate were used: TEC 8 conducting glass, supplied by Hartford Glass Co., Inc. (sheet resistance: 8 Ω/square) and ITO–60 conducting plastic, supplied by IST (60 Ω/square). After evaporation of the ethanol the substrate with the attached powder film was put between two planar steel press plates and pressure was applied by using a hydraulic press. Alternatively, deposition of a nanostructured TiO_2 film on a flexible conducting plastic substrate was performed using a roller mill (supplied by Oy Gradek AB in Finland) with two co-operative rollers. The compression was performed by allowing the substrate with the deposited powder film to pass between the rollers with a speed of about 1m s^{-1} and with a roller line pressure of 400 kN m^{-1}. Films were characterized by profilometry (Dektak 3), scanning electron microscopy, N_2-BET adsorption and X-ray diffraction.

Dye-adsorption and Electrolyte Preparation

The TiO_2 electrodes were stained to a dark red color by immersion for 2 hours in a dye bath consisting of an ethanolic solution of 0.5 mM cis-bis(isothiocyanato)bis(2,2´bipyridyl-4,4´dicarboxylato) ruthenium (II) bis-tetrabutylammonium (N719, purchased from Solaronix S.A., Switzerland). Two different electrolytes were used; the first electrolyte consisted of 0.5 M LiI, 0.05 M I_2 and 0.5 M t-butyl pyridine (all from Aldrich) in 3-methoxypropionitrile (Fluka) and was used at high light intensities (100-1000 W m^{-2}), the second electrolyte, used for low intensities (230 lux, fluorescent lamps), was identical except that the iodine concentration was reduced to 0.01 M.

Photoelectrochemical Measurements

The i-v characteristics of dye-sensitized nanostructured electrodes were measured in a sandwich electrode configuration. The counter electrode consisted of a thermally platinized conducting glass (6). The i-v curves were recorded using a computerized Keithley 2400 source meter. A sulfur lamp (Fusion Lightning Lightdrive 1000) was used to simulate sunlight. The light intensity was measured using a calibrated pyranometer (Kipp & Zonen CM11). In order to account for the spectral mismatch of the light source compared to natural sunlight, a conversion factor was calculated by comparing measurements in the system and in direct sunlight. Monochromatic incident photon-to-current conversion efficiencies were measured in a computerized set-up consisting of a xenon lamp, a monochromator with appropriate filters and lenses, Keithley 2400 source meter and a calibrated power meter (Newton 1830-C).

Results

Characterization of TiO_2 Powder and Compressed Films

The TiO_2 powder used in this study, Degussa P25, consists of highly agglomerated nanoparticles of anatase and rutile. XRD revealed that the composition was 70 % anatase and 30% rutile, with particle sizes of 22 nm and 35 nm, respectively. Before compression, the powder films are white and highly light scattering due to the size of the aggregates. After compression with pressures exceeding 200 kg cm^{-2}, films appear more blueish-white and are translucent for yellow-orange light. This is a clear indication that aggregates are broken into smaller units during compression. Further confirmation comes from the fact that the specific surface area increases after compression. With an applied pressure of 1000 kg cm^{-2} the BET surface area of the film was 63.9 m^{2}g^{-1} compared to 60.6 m^{2}g^{-1} for the powder. Both samples were annealed for 30 minutes at 450°C before the BET adsorption experiment.

The BET pore size distribution in the compressed film ranged from 5 nm up to 60 nm with a maximum at 23 nm and an average pore diameter of 23 nm. Thus, the pores in the compressed film have approximately the same diameter as the particle size. The average pore diameter of the un-pressed powder was 27 nm, however, the pore size distribution was much broader, ranging from 5 nm and up to 200 nm having a small and broad maximum at 30 - 40 nm.

The porosity of compressed films was determined by gravimetric analysis. It is a function of the applied pressure, see Figure 2. The porosity varies from 69% at 200 kg cm^{-2}, which is the lower limit to achieve mechanically stable and adherent films, to less than 50% at pressures exceeding 2000 kg cm^{-2}. For comparison, the porosity of screen-printed P25 films was reported to be 52% (5).

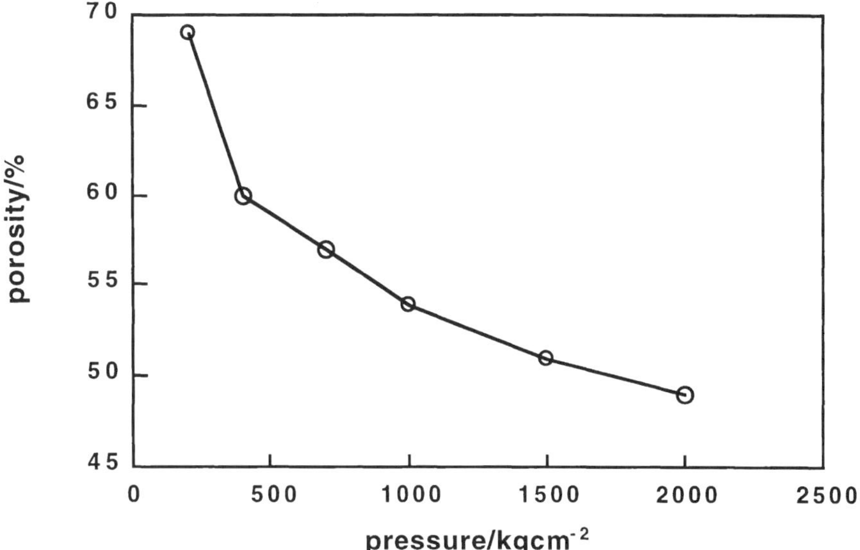

Figure 2. Porosity of compressed TiO$_2$ powder films as function of applied pressure.

128

Solar Cell Measurements

Conducting Glass Substrates

Spectra of the incident photon-to-current conversion efficiency of N719-sensitized pressed TiO_2 films with different thickness are shown in Figure 3. The spectrum roughly corresponds to the absorption spectrum of the dye. N719-dye is characterized by a dark red color and has a very broad spectrum. The extinction coefficient, however, is very low in the red and near infrared region. The sensitized nanostructured TiO_2 film has to be sufficiently thick to absorb a significant part of the sunlight in this region. The systematic increase in the IPCE at 700 nm is attributed to increased light absorption. In thicker films, however, more problems with diffusion limitation are encountered and the photovoltage tends to be lower, as the dark current contribution is increased. We are therefore currently developing films with enhanced light scattering properties by addition of larger TiO_2 particles.

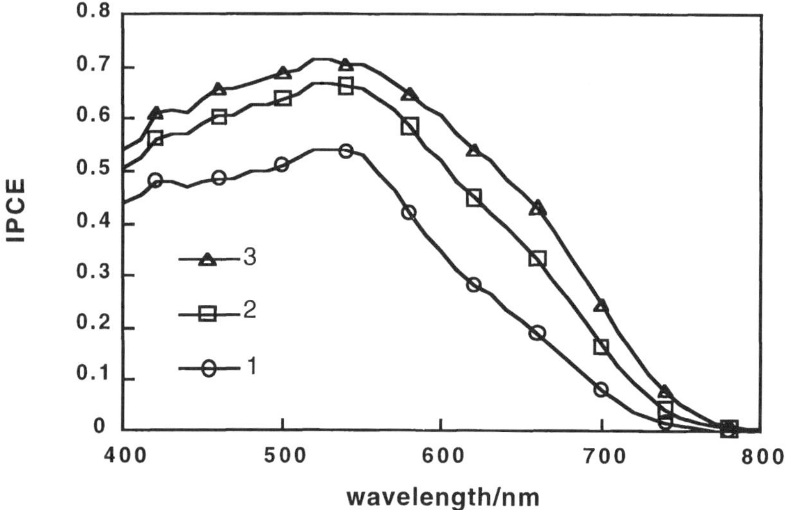

Figure 3. IPCE spectra of electrodes having stacked layers. Number of layers is indicated and each layer corresponds to 4 µm.

Tables I and II show the data extracted from i-v curve measurements on N719-sensitized nanostructured TiO_2 electrodes on conducting glass. The film thickness was kept constant at 8 μm. The manufacturing parameters for all electrodes were kept constant with the only exception that electrodes in Table I were not sintered before dye adsorption, and electrodes in Table II were sintered at 450°C for half an hour before dye adsorption. The fill factor is given by: FF= P_{max}/ (V_{oc}*I_{sc}).

Table I. Solar cell characteristics of compressed TiO$_2$ films without sintering*.

Light Intensity / W m^{-2}	Fill Factor	Efficiency/%	V_{oc}/V	I_{sc}/mAcm^{-2}
100	0.67±0.01	4.5±0.3	0.66±0.03	1.0±0.1
400	0.58±0.04	4.2±0.1	0.72±0.03	4.0±0.3
700	0.50±0.06	3.5±0.3	0.73±0.01	6.8±0.3
1000	0.47±0.06	3.0±0.4	0.73±0.01	8.6±0.1

*Film thickness: 8μm. Values are the average of 4 electrodes.

Table II. Solar cell characteristics of compressed TiO$_2$ films with sintering*.

Light Intensity / W m^{-2}	Fill Factor	Efficiency/%	V_{oc}/V	I_{sc}/mAcm^{-2}
100	0.64±0.01	4.3±0.3	0.64±0.01	1.0±0.1
400	0.57±0.02	4.1±0.2	0.69±0.01	4.1±0.3
700	0.52±0.03	3.5±0.2	0.71±0.01	6.6±0.5
1000	0.48±0.04	3.1±0.2	0.72±0.01	9.1±0.8

*Sintered at 450°C for 30 minutes. Film thickness: 8μm. Values are the average of 7 electrodes.

A comparison between the values in Tables I and II shows that the i-v characteristics of the sintered and the unsintered nanostructured TiO$_2$ electrodes are very similar. The linearity of the short circuit current with light intensity shows that the diffusion of the redox couple through the nanostructured film is not limiting for the photocurrent. Thus, the efficiency drop at higher light intensities can be attributed mainly to series resistance losses in the conducting substrate.

Conducting Plastic Substrates

The solar cell characteristics of pressed TiO$_2$ films on conducting plastic were measured in combination with a conducting plastic counterelectrode. It is necessary to improve the catalytic activity of the counterelectrode by

130

platinization. The usual thermal platinization requires a temperature of 385 °C
(*6*), which is incompatible with plastic substrates. However, by pressing carbon-
or platinized SnO_2 powder on conducting plastic it is possible to produce
counter electrodes at lower temperatures. The i-v curve for an all-plastic
sandwich cell is shown in Figure 4.

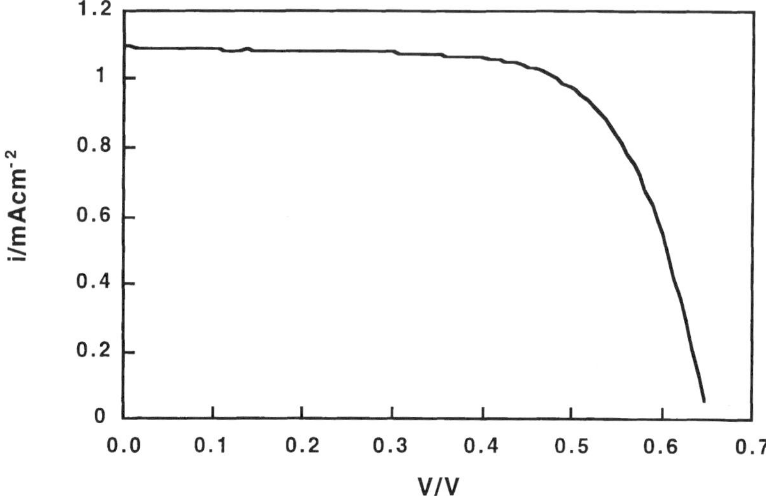

*Figure 4. Current-voltage curve of a plastic dye-sensitized solar cell. Light
intensity: 100 W/m², thickness TiO₂ film: 8.3 µm. The overall efficiency is 4.9 %.*

The overall cell efficiency of this cell, which was produced without heat
treatment and had a counter electrode of pressed platinized SnO_2, was 4.9 %.
The average efficiency for a large number of films on plastic (n = 200) was 4.0
± 0.5%. Because of the losses due to series resistance in the conducting plastic
layer (sheet resistance: 60 Ω/square) it was very difficult to reach high
conversion efficiencies with all-plastic dye-sensitized solar cells at higher light
intensities. Nevertheless, we achieved an efficiency of 2.3 % under 1 sun
illumination (1000 W m^{-2}).

At low-light indoor illumination conditions (230 lux) the performance of
plastic dye-sensitized solar cells is quite respectable, with a fillfactor, open-
circuit voltage and short circuit current of 0.59, 0.54 V and 18µAcm^{-2},
respectively. These results are comparable to those obtained for commercial
amorphous silicon cells and prototype screen-printed dye-sensitized cells on
conducting glass (*5*).

A brief study confirmed the possibility of continuous processing of
nanostructured TiO_2 on flexible substrates. Using a roller mill, a TiO_2 powder
film on conducting plastic was compressed at a speed of 1m s^{-1}. The resulting

nanostructured film was stable and showed promising solar cell performance after dye-sensitization. An average efficiency of 2.8 ± 0.2% was found at a light intensity of 100 W m^{-2}.

Discussion

By compression of a TiO_2 powder film a mechanically stable nanostructured film is formed. The powder film itself can be easily wiped off, whereas the pressed film sticks firmly. These characteristics allow for very precise patterning of the deposited film by means of selective pressing. The good solar cell characteristics reveal that conduction of photoinjected electron through the pressed nanostructured TiO_2 film is not a problem. In other words, there is a good electrical contact between the individual particles.

During the compression step very large forces occur locally, which apparently lead to efficient connection of the particles. The similarity of compressed films and fired films suggests that the particles are sintered together in both cases. A detailed high-resolution TEM study is currently under way to prove this assumption.

It is very advantageous that the firing step can be omitted. First, it is usually a time and energy-consuming step. Second, dye-adsorption, another time consuming step in the production of dye-sensitized solar cells, can in principle now be done on the powder before pressing. Initial experiments with pre-adsorption of the dye on TiO_2 powder gave promising results. Third, flexible plastic substrates, which can not withstand high temperatures, can now be used. This allows in principle for continuous roll-to-roll processing of dye-sensitized solar cells with a very high throughput and at a potentially low cost.

In conventional methods for preparing nanostructured films for dye-sensitized solar cells a firing step to ~450°C is required to obtain good electric contact between the particles. In principle, functioning dye-sensitized solar cells can be made using a low-temperature sintering/drying step (100°C for 24 h) to form a stable film from a colloidal TiO_2 paste (7). The quantum efficiencies for photocurrent generation achieved in this manner are below 50% (compared to close 100% for fired films), indicating losses due to poor electronic contact between the particles. The main disadvantage for a production viewpoint, however, is the slow manufacturing procedure.

Conclusion

Nanostructured TiO_2 films have succesfully been manufactured at room temperature of by means of compression of a powder film. The performance of these films in dye-sensitized solar cells is good: overall solar cell efficiencies exceeding 4% have been obtained for films deposited on conducting glass and conducting plastic. The presented technique allows for high-speed production of nanostructured films at room temperature.

Acknowledgements

The work in this paper has been financed the Swedish Foundation for Strategic Environmental Research (MISTRA) and the Swedish National Energy Administration (SEM).

References

1. O'Regan, B.; Grätzel, M. *Nature (London)* **1991**, *353*, 737-740.
2. Nazeeruddin, M. K.; Kay, A.; Rodicio, I.; Humphry, B. R.; Müller, E.; Liska, P.; Vlachopoulos, N.; Grätzel, M. *J. Am. Chem. Soc.* **1993**, *115*, 6382-6390.
3. Nazeeruddin, M. K.; Péchy, P.; Renouard, T.; Zakeeruddin, S. M.; Humphry-Baker, R.; Comte, P.; Liska, P.; Cevey, L.; Emiliana Costa; Shklover, V.; Spiccia, L.; Deacon, G. B.; Bignozzi, C. A.; Grätzel, M. *J. Am. Chem. Soc.* **2001**, *123*, 1613 -1624.
4. Barbe, C. J.; Arendse, F.; Comte, P.; Jirousek, M.; Lenzmann, F.; Shklover, V.; Grätzel, M. *J. Am. Ceram. Soc* **1997**, *80*, 3157-3171.
5. Burnside, S.; Winkel, S.; Brooks, K.; Shklover, V.; Grätzel, M.; Hinsch, A.; Kinderman, R.; Bradbury, C.; Hagfeldt, A.; Pettersson, H. *Journal of Materials Science-Materials in Electronics* **2000**, *11*, 355-362.
6. Papageorgiou, N.; Maier, W. F.; Grätzel, M. *J. Electrochem. Soc.* **1997**, *144*, 876-884.
7. Pichot, F.; Pitts, J. R.; Gregg, B. A. *Langmuir* **2000**, *16*, 5626 -5630.

Chapter 11

Indium–Tin Oxide Organic Interfaces

C. L. Donley[1], D. R. Dunphy[1,2], W. J. Doherty[1], R. A. P. Zangmeister[1],
A. S. Drager[1], D. F. O'Brien[1], S. S. Saavedra[1], and N. R. Armstrong[1,*]

[1]Department of Chemistry, University of Arizona, Tucson, AZ 85721
[2]Current address: Advanced Materials Laboratory, Sandia National Laboratories,
Albuquerque, NM 87185

We show here the detailed surface characterization of indium-tin oxide (ITO) thin films, and the effect of variations in surface composition on the electron transfer rates of a chemisorbed probe molecule, ferrocene dicarboxylic acid $(Fc(COOH)_2$. Pretreatment schemes vary both the extent of hydroxylation of the ITO surface, and the relative concentration of oxide defects, which help to control electron transfer events at the ITO interface. In addition we show the first example of spectroelectrochemical characterization of charge injection processes at ITO interfaces with ultra-thin films of new discotic mesophase (phthalocyanine) molecular assemblies, using broad-band visible wavelength spectroelectrochemistry. Prism coupling schemes with ITO attenuated total reflectance (ATR) elements, and CCD detection, allow for simultaneous characterization of the 450-900 nm region during oxidative doping of these assemblies.

Introduction

Indium-tin oxide (ITO) is a transparent conductive oxide that is widely used as a "passive electrode" in liquid crystal displays, and is seeing increasing use as an "active electrode" in organic light emitting diode displays (OLEDs) and photovoltaic (PV) cells, requiring passage of high current densities (1-5). Its use as a working electrode in several electroanalytical schemes over the last two decades is also well documented (6, and references therein). The performance of these technologies, however, is critically dependent upon high rates of charge transfer, and good chemical compatibility between the ITO surface and individual molecules, or molecular assemblies.

We have recently shown that integrated optical waveguides can be overcoated with 25-50 nm ITO films (creating an electroactive IOW or "EA-IOW") for spectroelectrochemical studies of extremely low coverages of adsorbed dyes, biomolecules, and molecular assemblies (6,7). As with ITO-coated attenuated-total-reflectance (ATR) elements, the evanescent field is most intense at the ITO interface, providing for spectroelectrochemical sensitivities, at a single wavelength in the EA-IOW, which are up to 10,000x those of a transmission spectroelectrochemical experiment, and 100-1000x those of ITO-ATR experiments (7).

Current electroactive waveguide technology is still limited in a number of ways. First, the outcoupled light intensity is <u>extremely</u> sensitive to changes in ITO surface composition, due to small changes in the refractive index of the ITO film that accompany compositional changes. These changes are thought to be linked to surface hydrolysis, leading to variations in surface hydroxide coverage, impurity incorporation, etc., which may also affect the adsorption of molecules to the ITO surface, and oxide defect densities which impact interfacial electron transfer rates. Our recent studies have suggested that these compositional changes are related to those which also limit the use of ITO anodes in both OLED and PV technologies. Understanding and controlling the composition of this surface is therefore critical to the use of this technology.

Another limitation of IOW or ATR spectroelectrochemical technology lies in the fact that only spectrometric characterization has been conducted to date, since coupling of more than one wavelength of light at a time into the ITO optical element has been problematic. Recent studies have suggested *i)* that it would be possible to (broadband) couple most of the visible wavelength spectrum into either an IOW or an EA-IOW (8), and *ii)* that broadband coupling of the entire visible spectrum into simple attenuated-total-reflectance (ATR) electroactive ITO/glass elements would be straightforward and could provide adequate sensitivity and high information content, in certain spectroelectrochemical experiments.

Figure 1 shows a schematic of this new ATR/waveguide spectroelectrochemical technology and some of the molecular systems which are candidates for study using sensitive broadband spectroelectrochemistry. The ATR technology is quite similar to previously reported ATR spectroelectrochemical systems (9) but now includes a collimated white light source, and the ability to disperse the out-coupled light via a diffraction grating, imaged onto a CCD camera detector, so as to provide for characterization of most of the visible wavelength region. Adsorbed dyes such as methylene blue (6,7) and bio-molecules such as cytochrome c (6) have been studied at coverages ranging from 1%-50% of a monolayer, where the absorbance changes undergone during reduction/oxidation allow for: *i)* reconstruction of a voltammetric response (absorpto-voltammetry (9) where the optical changes follow the Faradaic current response), *ii)* measurement of absorbance changes at both TE and TM polarizations which provide for an estimation of orientational changes occurring during redox processes.

Molecular assemblies based on discotic mesophase materials have also been studied using these technologies, where the disk-like monomer has been a phthalocyanine (Pc) (10,11). Functionalized silicon phthalocyanines which are rigid-rod polymers (linked through the O-Si-O spine), and linear aggregates of metallated and non-metallated Pcs, as sub-monolayer coverage thin films, show comparable spectroelectrochemical activity. Spectroelectrochemical studies of these molecular assemblies have previously suffered from the fact that the EA-IOW had so much sensitivity that full monolayer films could not be explored (due to extensive optical losses). In addition, only single wavelengths could be explored with each voltammetric scan, and spectral overlap of the reactant and product species made full characterization of the redox processes problematic. In all of these studies, it was also observed that composition changes to the ITO surface caused background optical changes which had to be factored into each of the spectroelectrochemical responses. Transmission spectroelectrochemical studies, on the other hand, did not have adequate sensitivity to follow absorbance changes occurring to the Pc film during redox events. Clearly an intermediate technology is needed, between the waveguide and transmission spectroscopic approaches, and we show here the efficacy of using ATR geometries, with broad-band coupling and CCD detection schemes, to characterize these new materials.

In this paper we *1)* review some of our recent monochromatic X-ray photoelectron spectroscopic (XPS) surface analysis of ITO electrode surfaces, in order to gain insight into changes which can occur during various solution and gas-phase pretreatments, and *2)* show the first results of a new broadband ATR spectroelectrochemical cell that allows for spectroelectrochemical studies on ca. 5.6 nm thick films of compact bilayer Pc molecular assemblies. The surface analysis studies document the variability in surface hydroxide, and oxygen-defect content of the ITO surface, and the affect of these changes on electron transfer rates of probe molecules, while the broad-band ITO/ATR capabilities prove to have good sensitivity for certain adsorbed redox-active thin films, such

Broadband ATR Technology

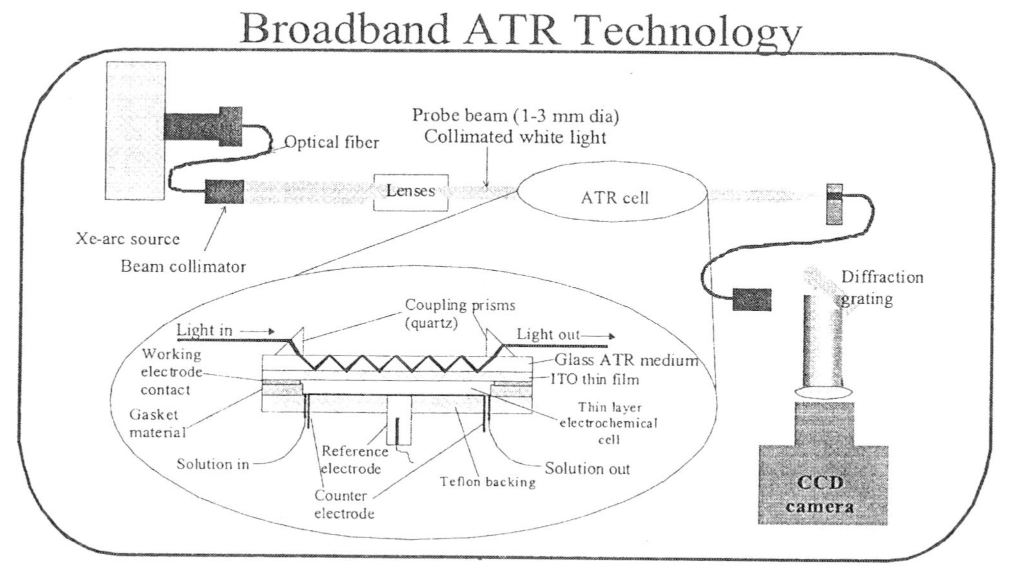

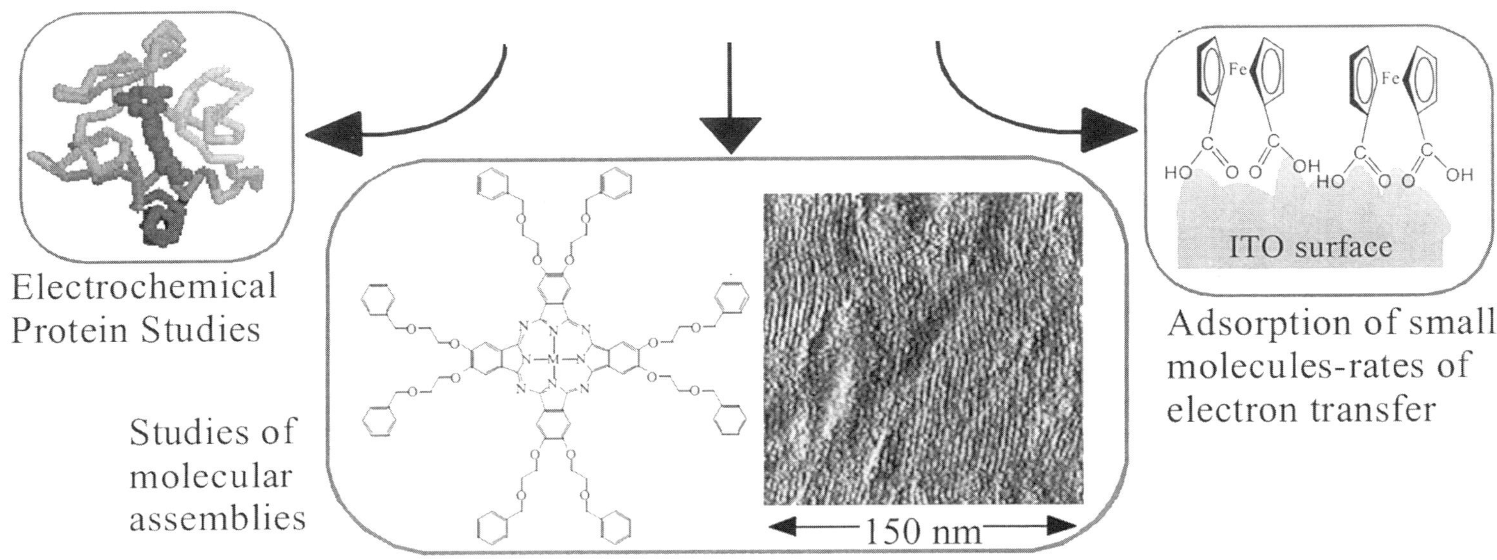

Figure 1. Top: A schematic of the ATR technology used for spectroelectrochemical studies. Bottom: Cytochrome C for protein adsorption studies, $CuPc(OC_2O)Bz_8$ molecular columns, $Fc(COOH)_2$ adsorbed to ITO.

as our new discotic mesophase phthalocyanines, and are less sensitive in their background optical changes than the IOW-based technologies.

Experimental

ITO Samples

ITO/glass used in this study was obtained from Donnelly Corp., with a sheet resistance of ca. 13 $\Omega/\square$. All ITO was first cleaned by scrubbing in a solution of Triton X-100 followed by successive sonication in Triton X-100, DI water, and ethanol for at least 10 minutes each. To ensure reproducibility all experiments described below were repeated numerous times, with separate ITO samples and all samples were examined (by XPS, electrochemical methods, etc.) both *before* and *after* pretreatments to serve as a reference.

XPS of Indium and Indium Oxide Standards

ITO surfaces were characterized by XPS using a Kratos Axis Ultra spectrometer (base pressure of ca. 10^{-9} torr, monochromatic Al Kα (1486.6 eV) excitation, with an analysis spot size of 300 x 700 microns, and a pass energy of 20 eV). Charge neutralization was used to compensate for charging effects with a charge balance voltage of 1.8v. These conditions were used after they showed a minimum FWHM and a maximum peak intensity, to ensure there was no peak broadening. These conditions also did not show any lineshape distortion relative to instances when the sample was in electrical contact with the spectrometer. In_2O_3 and $In(OH)_3$ standard powders were obtained from Fluka and Aldrich respectively, and the In_2O_3 was heated to 160°C for 2 hours in an effort to remove any hydroxide contamination. These powders were mounted in indium metal foil; the exposed metal portion then also served as a reference for XPS binding energy correction (In ($3d_{5/2}$) peak at 443.2 eV).

The "InOOH" standard is an intermediate oxy-hydroxide that was electrochemically grown on a clean indium foil following a method developed by Metikos-Hukovic and Omanovic (12). After analysis of this sample by XPS, the surface was sputtered with an argon ion beam until the indium metal peak was observable by XPS which allowed for this peak to be used for binding energy correction.

Pretreatment of the ITO Surface

Several pretreatment protocols have been used with ITO surfaces which provide for removal of contaminants, increases of surface work function, improvements in OLED and PV device performance (13-16), and also improvements in electron transfer rates of chemisorbed molecules. A piranha treatment has been developed by Wilson and Schiffrin (17) which involves soaking the ITO in a 10 mM solution of NaOH for 4 hours at 80°C, followed by a one minute soak in piranha (4:1 solution of H_2SO_4:H_2O_2), and heating in air for 2 hours at ca. 160°C. Air plasma treatment involves plasma cleaning in a Harrick plasma cleaner (Model PDC-32G, 60 W) for 15 minutes. Finally, the RCA treatment consists of soaking the ITO in a 1:1:5 solution of NH_4OH : H_2O_2 : H_2O at 80°C for 30 minutes.

Ferrocene Dicarboxylic Acid Adsorption Studies

The adsorption of ferrocene dicarboxylic acid ($Fc(COOH)_2$) onto the ITO surface occurred by soaking the ITO in a 1 mM solution of the ferrocene in ethanol for 10 minutes, and then rinsing with acetonitrile, as recently reported by Zotti and coworkers (18). Electrochemistry was carried out on $Fc(COOH)_2$ modified ITO in a 0.1 M solution of tetrabutylammonium hexafluorophosphate in acetonitrile. The counter electrode was a platinum wire, the reference electrode was a saturated Ag/AgCl electrode, and all potentials were referenced to that for the oxidation/reduction of ferrocene (Fc/Fc^+ couple).

ATR Spectroelectrochemistry of $CuPc(OC_2OBz)_8$

Bilayer films of 2,3,9,10,16,17,23,24- octakis-(2-benzyloxyethoxy) phthalocyaninato copper ($CuPc(OC_2OBz)_8$) were prepared using the Langmuir-Blodgett technique, with films transferred to ITO substrates by horizontal transfer (Schaefer method). These bilayer films show unique stability allowing for nearly complete transfer from the trough to the substrate (10). These films on ITO were then immersed in 0.1M $LiClO_4$ in water in a cell similar to the schematic shown in Figure 1. The reference electrode was a saturated Ag/AgCl electrode, the counter electrode was platinum wire. The scan rate was 5 millivolts/sec, and spectra were recorded every 10 seconds. For these experiments, only half of the ITO slide was covered with the $CuPc(OC_2OBz)_8$ bilayer film, the other half served as a blank, and by adjusting the position of the light travelling through the glass substrate, either the blank side or the side with Pc could be addressed. Blank spectra were taken at every potential that sample spectra were taken and were subtracted from the sample spectra.

Discussion

XPS of Indium Standards and As-Received ITO

The current state of understanding of ITO thin films (19) and our XPS, electrochemical and AFM characterization of these surfaces leads to the formulation of a complex model of the ITO surface shown in Figure 2. Early studies of ITO surfaces by XPS showed slightly asymmetric In (3d), Sn (3d), and O(1s) peaks suggesting the presence of only one form of indium and tin at the surface, and no more than two forms of oxygen (20,21). Monochromatic X-ray sources provide higher resolution photoemission spectra indicating the presence of up to 2-3 indium and tin species and up to four oxygen species on the ITO surface. Our XPS characterization studies, comparing ITO films with indium standard spectra (Figure 3), suggest that the ITO lattice is terminated with "InOOH" species, and regions of the fully hydroxylated indium oxide, $In(OH)_3$, are also present. Previous studies of hydrolysis of indium oxides suggests that the $In(OH)_3$ species is not incorporated in the ITO lattice, but that clusters of this species can remain at the surface, in part due to the rather low solubility of $In(OH)_3$ (22). In commercial ITO the In_2O_3 bixbyite lattice is doped with Sn^{+4} and possesses a significant fraction of oxygen deficiency sites added to improve the electrical and optical properties of the film (19). In addition, it is apparent in the XPS data that there are specific O(1s) peaks due to oxygen atoms which are proximate to oxygen defect sites and they appear at binding energies intermediate between those for In_2O_3 oxygen, and those seen for InOOH and $In(OH)_3$ (1,20).

When these ITO surfaces were characterized at high photoemission take-off angles, to accentuate the near surface composition for both the In(3d) and the O(1s) peaks, the intensity of the higher binding energy components increased, indicating a larger presence of these hydroxide, oxy-hydroxide, and oxide defect species in the near surface region. Given the roughness seen in the AFM images of these surfaces, with features dominated by columnar structures growing from the bulk (Figure 2), this is not a surprising result.

Pretreatment of the ITO Surface

Several ITO surface pretreatments have been developed to change the ITO workfunction for use in "active" devices (13-16). In an effort to evaluate these different protocols for cleaning the ITO surface, and preparing it for subsequent chemical modification, a series of these pretreatments were evaluated, including a piranha/heat treatment (originally developed by Schiffrin and coworkers,

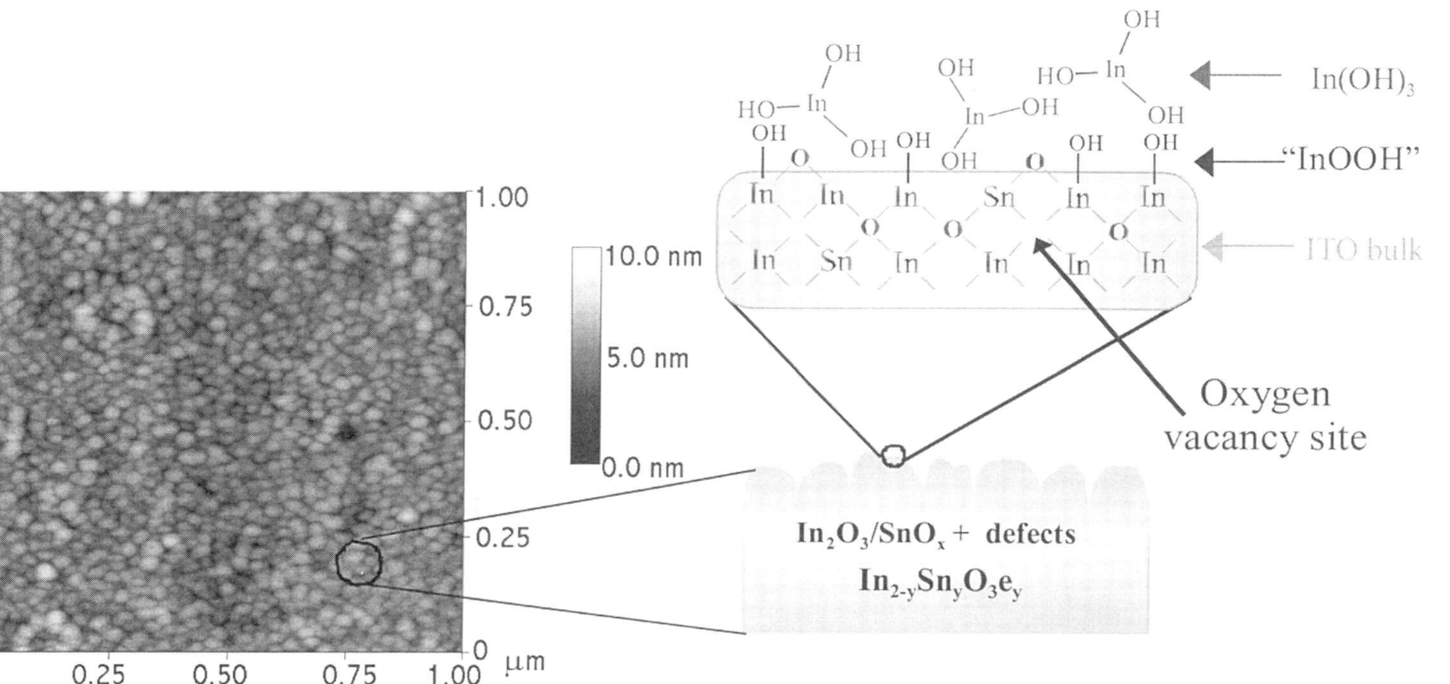

Figure 2. Model of the ITO surface.

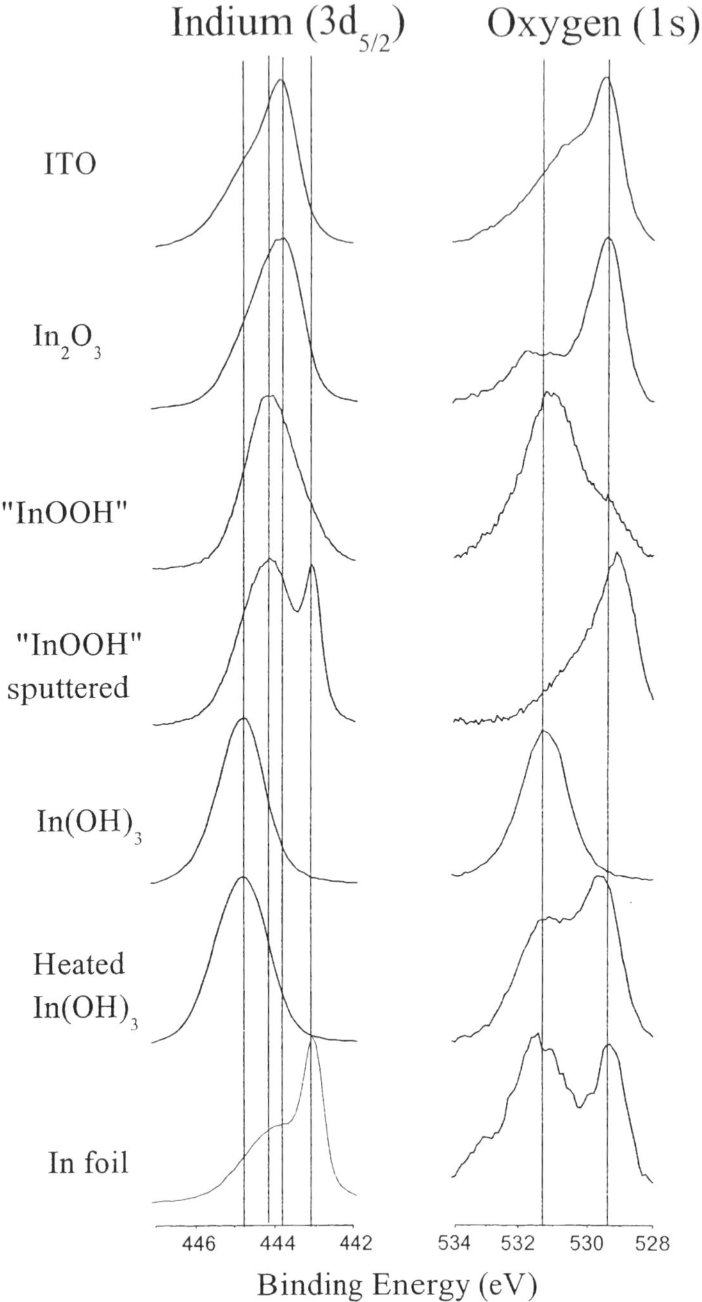

Figure 3. In(3d_{5/2}) and O (1s) XPS peaks for ITO and a series of indium standards.

ostensibly to allow for high coverages of silanes to be attached to the ITO surface) (17), an air plasma treatment (a treatment which appears to produce the best response in OLEDs based on ITO), and an RCA treatment. While we did not observe large changes in either the In(3d) or Sn(3d) peaks following pretreatment, we did see characteristic changes to the O(1s) peak taken at a 75° take-off angle, which we focus on in Figure 4. To ensure that the *changes* in the ITO surface with pretreatment were being monitored, and not just variations in different pieces of ITO, each ITO sample was examined by XPS both *before* and *after* pretreatment.

Figures 4a, c and e are typical O(1s) peaks for as-received (detergent, water, ethanol cleaned) ITO. A brief explanation of the peak assignments is necessary. Peaks A and C fall at the same energy positions as that for the O(1s) peaks in In_2O_3 and $In(OH)_3$ standards. Peak B is assigned to photoemission from oxygen atoms nearest to defect sites in the oxide lattice, which donate some of their electron density towards the indium atoms that are no longer fully coordinated, to compensate for the lost oxygen (1,20). This donation of electron density towards the indium appears to cause the O(1s) peak to shift towards higher binding energy and at the same time partially compensates the indium for the loss of adjacent oxygen. Large chemical shifts in the In(3d) spectra are therefore not seen due to the formation of these oxygen defects.

Figures 4a and b illustrate the changes in the O(1s) lineshape following piranha treatment of the ITO surface. The important features to notice are the large increase in the surface hydroxide concentration (peak C), and the slight decrease in the number of oxygen defect sites. The effects of air plasma cleaning are shown in Figures 4c and d. Although the effect is not as dramatic as for the piranha treatment, there is an increase in surface hydroxides. Finally, Figures 4e and f show that the RCA treatment slightly reduces both the amount of surface hydroxide and the relative amount of near-surface oxygen defects. These treatments demonstrate the range of hydroxide and oxygen vacancy concentrations obtainable at the ITO surface.

Adsorption of Ferrocene Dicarboxylic Acid to ITO

To evaluate the pretreatments of the ITO surface in terms of the electron transfer rates there, $Fc(COOH)_2$ was used as a probe molecule and adsorbed to the surface at maximal coverage. The cyclic voltammograms for $Fc(COOH)_2$ adsorbed to various ITO surfaces are shown in Figure 5. These voltammograms represent the electrochemical activity seen on the first voltammetric sweep, subsequent scans showed the loss of ca. 20% of the adsorbed material, and a shift of the remaining voltammetric response to less positive potentials. Different background currents seen at potentials negative of the redox process suggest some changes in surface area of the ITO film. From background-

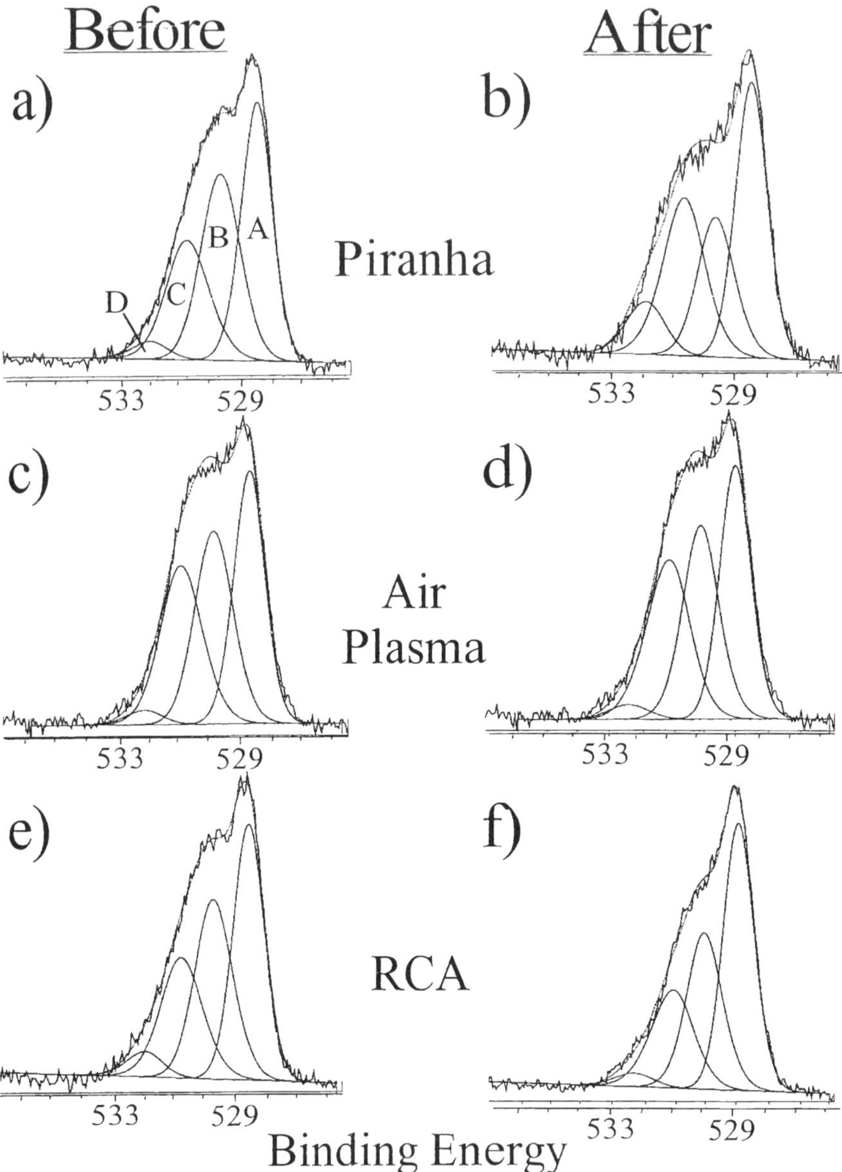

Figure 4. O (1s) XPS data taken at a 75° take-off angle following ITO surface pretreatment.

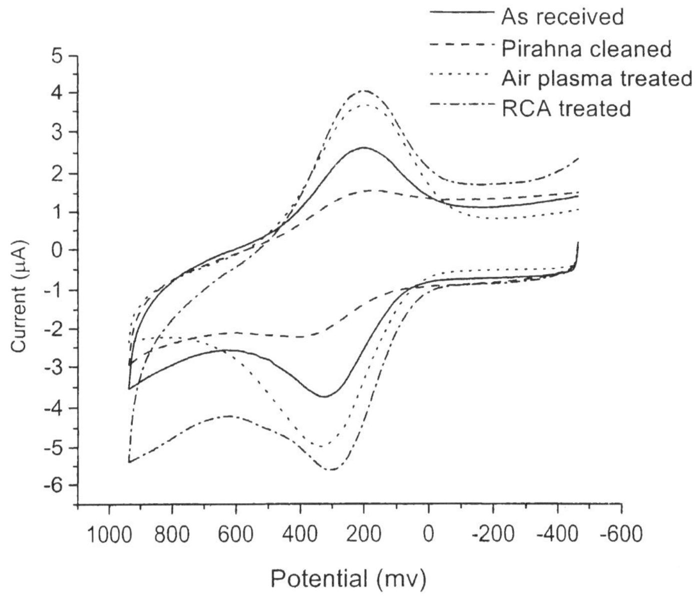

Figure 5. Cyclic voltammetry for Fc(COOH)$_2$ adsorbed to pretreated ITO.

corrected i/V curves it is clear that the various pretreatments give different coverages of Fc(COOH)$_2$ and significantly different electron transfer rates. These results are summarized in Table 1.

Table 1. Summary of XPS and Electrochemical Data on Pretreated ITO

Treatment	Relative Hydroxide Conc.[a]	Adsorbed Ferrocene (*10^{-10} moles/cm^2) (% ML)	Electron Transfer Rate (s^{-1})	Redox Potential (mv)[b]
Plain	3rd	0.7 ± 0.3 (17%)	0.7 ± 0.2	264
Piranha	1st	0.2 ± 0.1 (5%)	0.167 ± 0.006	302
Air Plasma	2nd	1.8 ± 0.9 (44%)	0.7 ± 0.2	263
RCA	4th	1.3 ± 0.4 (32%)	0.8 ± 0.2	257

[a] As determined by XPS analysis
[b] potentials vs. Fc/Fc$^+$ solution redox couple in MeCN

It is clear that the piranha treatment gives the lowest coverage of Fc(COOH)$_2$ (even though it gives the highest surface hydroxide concentration by XPS), and the slowest electron transfer rates. Our XPS results show that the increase in hydroxide concentration following piranha treatment could be due to either "InOOH" or In(OH)$_3$, and these species are not removed by insertion of the ITO sample into the vacuum environment. We proposed that this solution treatment creates an excess of In(OH)$_3$ at the surface which is not tightly bound (incorporated) into the main oxide lattice. While these excess hydroxides are available both for silane attachment and chemisorption of Fc(COOH)$_2$, it is likely they can be dislodged from the surface to reduce the effective concentration of bound redox couple, and in addition, may not be in good electrical contact with the ITO bulk which may lead to slow electron transfer rates. The RCA treatment produces the highest electron transfer rates observed for the Fc(COOH)$_2$ chemisorbed species, but only an intermediate Fc(COOH)$_2$ coverage, and the lowest surface hydroxide coverage.

The air plasma treatment yielded the highest coverage of Fc(COOH)$_2$ on ITO, but the O(1s) peak shapes alone do not account for this difference. A closer examination of the In(3d) peak for these samples showed that a best-fit of these peaks required a contribution from an "InOOH" species, (Figure 3) in addition to the In$_2$O$_3$ and In(OH)$_3$ components. As shown schematically in Figure 2 the air plasma treatment appears to produce a significant coverage of the hydroxides that terminate the ITO lattice and which we hypothesize are more effective in chemisorbing electroactive Fc(COOH)$_2$ vs. In(OH)$_3$ simply adsorbed to the ITO surface. These studies are of course complicated by the fact that the XPS analysis is done in high vacuum conditions, while the adsorption and redox chemistry of the adsorbed probe molecules are done in solvents designed to promote hydrogen bonding of the carboxylic acid functionalities to the hydroxylated surface, but which may promote a surface composition which may be slightly different to that found in vacuum. Clearly *in-situ* probes of surface composition for the ITO surface would be useful in further characterizing these processes.

ATR Spectroelectrochemistry of Phthalocyanines

The spectroelectrochemistry of one bilayer of $CuPc(OC_2OBz)_8$ transferred to ITO by the Langmuir Blodgett technique was carried out in the ATR geometry with ca. 12 reflections. This Pc has shown the remarkable ability to form coherent linear aggregates, over 100 nm in length, with the individual Pc monomers sitting edge-on in the column, tilted away from the column axis in a configuration reminiscent of the staggering seen in β-phase CuPc (23). Unusually stiff, 5.6 nm thick bilayer films can be transferred to a variety of substrates for further characterization. These aggregate assemblies show strong optical and electrical anisotropies, with conductivities along the column axis at least a factor of 3-4x larger than those across the column (23). Their use in both organic field effect transistor and organic photovoltaic technologies is being explored currently.

The voltammetric data for the oxidation of the single bilayer film of this material is shown in Figure 6a, while the spectroelectrochemical data is shown in Figure 6c and 6d. Previous studies of such self-assembled Pc thin films (10), and comparable thin films created from silicon phthalocyanine polymers (24,25), have shown that their oxidation in contact with aqueous media occurs in two potential regions, represented here by the well-defined voltammetric peaks (1) and (2). Charge transfer events for this single bilayer film appear to have high rates (i.e. rapid equilibration of the Pc film to changes in electrode potential are possible), and are quite reversible. For thicker Pc films these voltammetric regions are not as well resolved as shown in Figure 6, and the unambiguous spectral characterization of the redox events has not been straightforward (10). Coulometric analysis of these peaks shows that ca. $6.3*10^{-6}$ coul/cm^2 of charge are transferred in the first oxidation wave, corresponding to the one-electron oxidation of ca. 20% of the Pc monomers in this bilayer film (assuming an area per molecule of $5.5*10^{-15}$ cm^2 which would give ca. $3.1*10^{-5}$ coul/cm^2 per monolayer of Pc). Assuming that peak (2) also corresponds to a one-electron oxidation event, the coulometric analysis yields $1.2*10^{-5}$ coul/cm^2, corresponding to an additional 40% of Pc monomer oxidation in these films. The summation of charge in both oxidation peaks therefore corresponds to the oxidation of ca. 60% of the Pc monomers in the bilayer film.

Previous studies have suggested that these two redox processes are not independent of each other, but rather represent the oxidation of the most

148

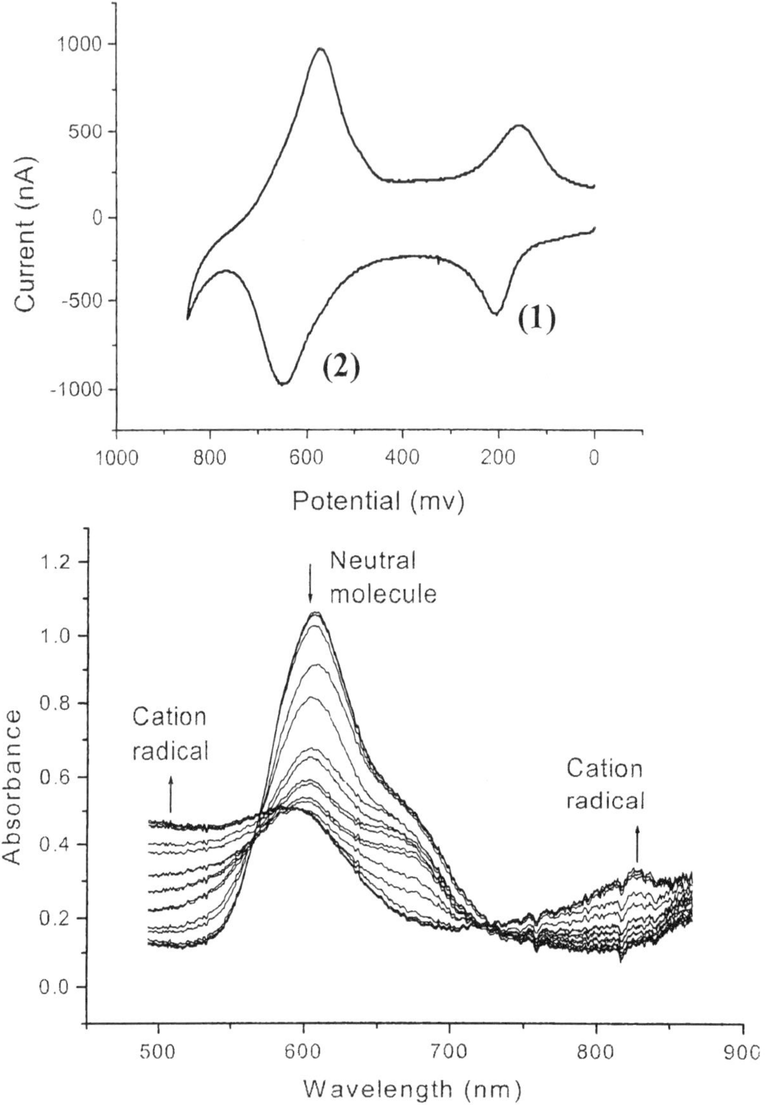

Figure 6. (a) Cyclic voltammetry to accompany the spectroelectrochemical data of CuPc(OC$_2$O)Bz$_8$ shown in (c). (b) Schematic of the CuPc(OC$_2$O)Bz$_8$ columns and the two sites for counter ion penetration. (d) Voltammetry recreated from absorbance data.

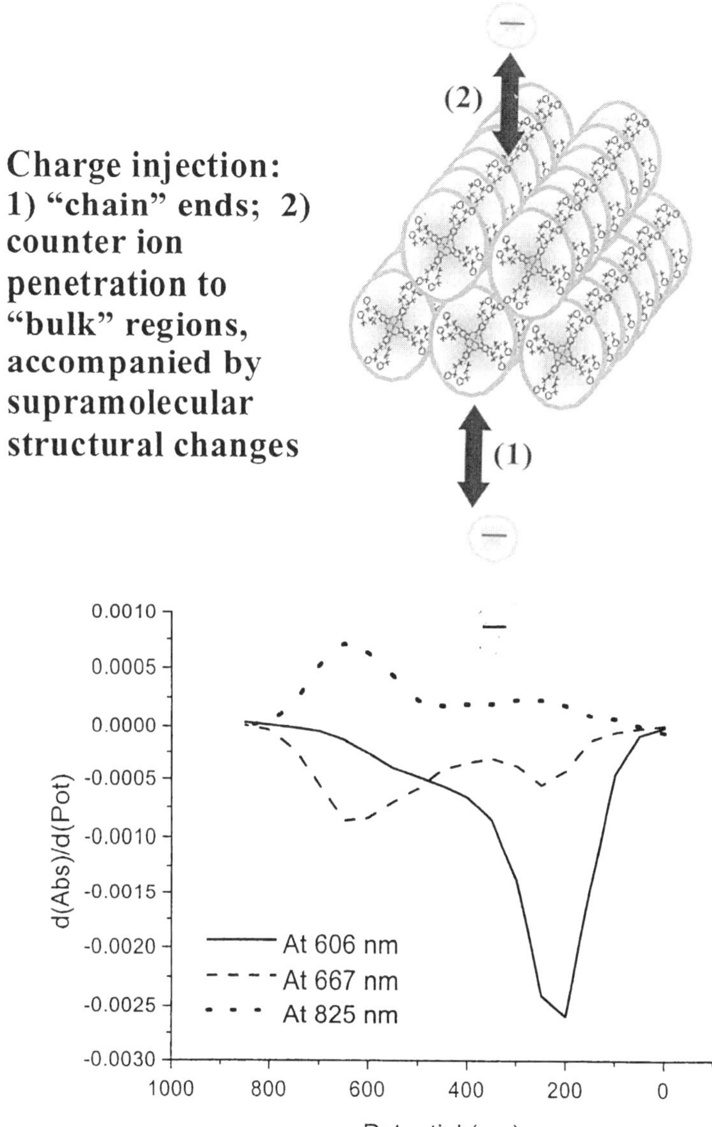

Figure 6. *Continued.*

accessible Pcs in each aggregate chain (peak (1)), and the further oxidation of Pc's which are in the middle of coherent aggregates, where counter ion penetration to achieve electroneutrality requires higher field strengths, as seen with many oxidizable conductive polymer systems (26 and references therein). We find that the relative sizes of these voltammetric peaks are strongly dependent upon sweep rate, counter ion size and charge density, and the degree of ordering in the Pc film, consistent with this hypothesis (10,11).

Previous transmission spectroelectrochemical characterization of these oxidation events, in thicker films suggested that the product of the Pc oxidation, the cation radical form, would show new absorbance features at ca. 820 nm, and at ca. 400-500 nm (10). Such absorbance features are confirmed in the spectra of Figure 6c, and the cation radical absorbance in the near-IR region is particularly well resolved. These new spectral features grow in as the absorbance due to the neutral Pc is reduced in intensity, but it is clear that the significant overlap of the absorbance bands below 750 nm can complicate the interpretation of the redox process – i.e. the optical properties of the remaining neutral Pc aggregate are affected by the presence of oxidized monomers within the Pc aggregate column.

Monitoring the absorbance changes at individual wavelengths, and then computing the first derivative of that absorbance change with respect to potential (dA/dE) produces an "absorptovoltammogram" (Figure 6d) whose magnitude at each wavelength is directly proportional to the faradaic current corresponding to that redox event, minus the capacitive current response, which does not produce an optical change in the thin film (6,7). The absorptovoltammetric response at 825 nm clearly shows the increase of the cation radical concentration in the Pc aggregate, in two stages, with most of the increase occurring during the second voltammetric wave. The loss of absorbance at 606 nm, corresponding to the neutral parent aggregate, is most pronounced in the first oxidation process. The absorbance change at 667 nm demonstrates the complexity of the absorbance bands left after partial oxidation of the Pc aggregate, showing its largest change during the final oxidation event.

These data suggest that slight structural changes may be occurring during oxidation of these Pc films, which change both the tilt angle of the monomer units within the linear aggregate, and the spacing between them, owing to the need to accommodate counter ions near the Pc cation radical, to achieve electroneutrality (10,11). Studies are underway with this spectroelectrochemical technology, using both TE and TM polarization during the above experiments, to determine whether changes in optical anisotropy occur during oxidation, and the degree to which the tilt angle of these Pc monomers is affected by the oxidation process. It should be noted that these oxidation events have been used to "dope" arrays of these Pc aggregates in microcircuit conductivity experiments, achieving increases in dark conductivity up to two orders of magnitude by oxidation at potentials of ca. 1.0 volts (27). Whether selective doping and adequate conductivities can be achieved in these linear arrays as to make them attractive as interconnects in organic circuitry is under exploration.

Conclusions

It is clear that optimization of ITO surfaces both for device applications and for electroanalytical characterization of adsorbed/covalently attached molecules requires significant spectroscopic characterization of the near surface composition of these materials, and a systematic correlation of composition changes with changes in electrochemical properties. The ITO surface has to be considered to be composed of stoichiometric oxides, trace amounts of "InOOH-like" species, which terminate the ITO lattice, and possibly adsorbed $In(OH)_3$ which is not incorporated in the lattice. Each pretreatment we have tried to date produces variable concentrations of each of these species. Electrochemical studies of the ITO surface following $Fc(COOH)_2$ adsorption indicate that the amount of $Fc(COOH)_2$ adsorbed and the rate of electron transfer vary with pretreatment. Preliminary studies to create OLEDs based on these surface modifications, and ferrocene adsorption suggest that the presence of even 50% of a monolayer of such a redox probe has a measurable, positive effect on device performance.

Using our optimized pretreatment procedures we show that rapid electron transfer to Pc aggregate assemblies is possible at the ITO surface, and that broadband ATR spectroelectrochemistry now enables the study of a single bilayer of such materials. These results suggest that complicated orientational changes may be occurring during Pc oxidation in the aggregate, owing to the difficulty in counterion incorporation, and that further spectroelectrochemical studies, using polarized radiation sources, are needed to follow these changes.

Acknowledgments

This research was supported in part by grants from the National Science Foundation (Chemistry and DMR, NRA and DFOB) and from the National Institutes of Health (SSS and NRA).

References

1. Kim, J.S.; Ho, P.K.H.; Thomas, D.S.; Friend, R.H.; Cacialli, F.; Bao, G.-W.; Li, S.F.Y. *Chem. Phys. Lett.* **1999**, *315*, 307-312.
2. Kim, J.S.; Cacialli, F.; Granstrom, M.; Friend, R.H.; Johansson, N.; Salanek, W.R.; Daik, R.; Feast, W.J.J. *Synthetic Metals* **1999**, *101*, 111-112.

3. Kim, H.; Gilmore, C.M.; Pique, A.; Horwitz, J.S.; Mattoussi, H.; Murata, H.; Kafafi, Z.H.; Chrisey, D.B. *J. Appl. Phys.* **1999**, *86*, 6451-6461.

4. Mason, M.G.; Hung, L.S.; Tang, C.W.; Lee, S.T.; Wong, K.W.; Wang, M. *J. Appl. Phys.* **1999**, *86*, 1688-1692.

5. Brabec, C.J.; Sariciftci, N.S.; Hummelen, J.C. *Adv. Funct. Mater.* **2001**, *11*, 15-26.

6. Dunphy, D.R.; Mendes, S.; Saavedra, S.S.; Armstrong, N.R. In *Interfacial Electrochemistry: Theory, Experiment, and Applications.*; Wieckowski, A., Ed.; Marcel Dekker, Inc.: New York, NY, 1999; pp 513-525.

7. Dunphy, D.R.; Mendes, S.; Saavedra, S.S.; Armstrong, N.R. *Anal. Chem.* **1997**, *69*, 3086-3094.

8. Bradshaw, J.T.; Mendes, S.; Armstrong, N.R.; Saavedra, S.S. Manuscript in preparation.

9. Heineman, W.R.; Hawkridge, R. M.; Blount, H.N. In *Electroanalytical Chemistry*; Bard, A. J., Ed.; Marcel Dekker, Inc.: New York, 1984; Vol. 13, pp 1-113.

10. Smolenyak, P.E.; Osburn, E.J.; Chen, S.-Y.; Chau, L.-K.; O'Brien, D.F.; Armstrong, N.R. *Langmuir* **1997**, *13*, 6568-6576.

11. Smolenyak, P.E.; Peterson, R.A.; Dunphy, D.R.; Mendes, S.; Nebesny, K.W.; O'Brien, D.F.; Saavedra, S.S.; Armstrong, N.R. *Porphyrins and Phthalocyanines* **1999**, *3*, 620-633.

12. Metikos-Hukovic, M.; Omanovic, S. *J. Electroanal. Chem.* **1998**, *455*, 181-189.

13. Le, Q.-T.; Forsythe, E.W.; Nuesch, F.; Rothberg, L.J.; Yan, L.; Gao, Y. *Thin Solid Films* **2000**, *363*, 42-46.

14. Milliron, D.J.; Hill, I.G.; Shen, C.; Kahn, A.; Schwartz, J. *J. Appl. Phys.* **2000**, *87*, 572-576.

15. Nuesch, F.; Rothberg, L.J.; Forsythe, E.W.; Le, Q.T.; Gao, Y. *Appl. Phys. Lett.* **1999**, *74*, 880-882.

16. Kim, J.S.; Granstrom, M.; Friend, R.H.; Johansson, N.; Salaneck, W.R.; Daik, R. Feast, W.J.; Cacialli, F. *J. Appl. Phys.* **1998**, *84*, 6859-6870.

17. Wilson, R. Schiffrin, D.J. *Analyst* **1995**, *120*, 175-178.

18. Zotti, G.; Schiavon, G.; Zecchin, S.; Berlin, A.; Pagani, G. *Langmuir* **1998**, *14*, 1728-1733.

19. Hartnagel, H.L.; Dawar, A.L.; Jain, A.K.; Jagadish, C. In *Semiconducting Transparent Thin Films*; Institute of Physics Publishing: Bristol, 1995; pp 3-15, 175-187, 265-282.

20. Fan, J.C.C.; Goodenough, J.B. *J. Appl. Phys.* **1977**, *48*, 3524-3531.

21. Major, S. K., S.; Bhatnagar, M.; Chopra, K.L. *Appl. Phys. Lett.* **1986**, *49*, 394-396.

22. Baes, C. F.; Mesmer, R.E. In *The Hydrolysis of Cations*; John Wiley & Sons: New York, 1976; pp 319-328.

23. Smolenyak, P.; Peterson, R.; Nebesny, K.; Torker, M.,; O'Brien, D.F.; Armstrong, N.R. *J. Am. Chem. Soc.* **1999**, *121*, 8628-8636.

24. Ferencz, A.; Armstrong, N.R.; Wegner, G. *Macromolecules*, **1994**, *27*, 1517-1528.
25. Dunphy, D.; Smolenyak, P.; Rengel, H.; Mendes, S.; Saavedra, S.S.; O'Brien, D.F.; Wegner, G.; Armstrong, N.R. *Polym. Prepr.* **1998,** *39*, 723-724.
26. Doblhofer, K.; Rajeshwar, K. In *Handbook of Conducting Polymers*; Skotheim, R.A.; Elsenbaumer, R.L.; Reynolds, J.R. Eds.; Marcel Dekker, Inc.: New York, NY, 1998; pp 531-588.
27. Zangmeister, R.A.P. Work in progress.

Chapter 12

Molecules as Components in Electronic Devices: Supramolecular Coordination Compounds as Components in Devices

Carlo A. Bignozzi[1] and Gerald J. Meyer[2]

[1]Dipartimento di Chimica, dell Universita, Centro di Fotoreattivitae Catalisi CNR, Università di Ferrara, 44100 Ferrara, Italy
[2]Department of Chemistry, Johns Hopkins University, Baltimore, MD 21218

Some strategies for the design of supramolecular coordination compounds for applications as molecular components in devices are discussed. Supramolecular compounds that efficiently absorb light and promote interfacial electron transfer processes when bound to semiconductor surfaces are of specific interest. When utilized as components in photonic devices, these supramolecular compounds allow the conversion of light into an electrical response to be controlled at the molecular level. Supramolecular compounds that feature additional functions such as intramolecular electron transfer and intramolecular energy transfer (antennae effects) have been studied for their potential use as components in solar energy conversion devices.

Over the last ten years, our groups have been working in the field of supramolecular photochemistry at semiconductor interfaces. In particular, we have been interested in the design, synthesis and characterization of supramolecular coordination compounds, i.e. inorganic coordination compounds with suitable built-in light-induced and electron transfer functions, that can be integrated into semiconductor devices. The goal of this contribution is to use representative experimental results, mainly taken from the work of our group, to exemplify, a supramolecular approach to the sensitization of wide bandgap semiconductors to visible light. Real-world applications of this science already exist for solar energy conversion. However, these studies provide a general molecular approach for the controlled conversion of light into an electrical response and thus have potential applications in displays, sensors, photochromics, electrochromics, and other molecular photonic devices.

Supramolecular Approach to Dye Sensitization of Wide Bandgap Semiconductors

The sensitization of wide bandgap semiconductors to visible light with molecular chromophores is well known (*1*). The recent order of magnitude increase in solar energy conversion efficiencies from regenerative solar cells based on dye sensitized electrodes has renewed interest in sensitization processes, particularly with nanocrystalline semiconductor thin films (*2*). The photophysical and photochemical properties of supramolecular coordination compounds in fluid solution have been well reviewed (*3*). The use of supramolecular compounds for dye sensitization may provide fundamental insights into interfacial electron transfer process that would not be gained with simple molecular compounds. Supramolecular sensitizers can also lead to more efficient devices for solar energy conversion and other applications. A supramolecular species possesses in general the following attributes:

(i) the intrinsic properties of the molecular components are not significantly perturbed, and (ii) the properties of the supramolecular system are not a simple superposition of the properties of the molecular components, there is a supramolecular function (*3*). Upon substitution of one of the molecular components by a semiconductor, a hetero-supramolecular system is formed. In addition to interfacial electron transfer, heterosupramolecular systems have been designed to support intramolecular electron transfer or energy transfer functions. In order for these heterosupramolecular systems to perform these and other desired functions, several non-trivial issues must be addressed. The first is the molecular architecture at the semiconductor surface where the appropriate molecular components must be assembled with suitable connectors in the right sequence such that each step is thermodynamically allowed. In addition, delicate issues of a kinetic nature must be addressed. As a matter of fact, the kinetics of each of the electron transfer steps should be optimised, such that absorbed photons are quantitatively converted into an interfacial charge-separated states and hence an efficient electrical response.

The following discussion is divided into two sections that both involve heterosupramolecular systems based on Ru(II) polypyridyl compounds bound to TiO_2 nanocrystallites (anatase) thin films. The first section describes assemblies designed to promote intramolecular and interfacial electron transfer upon light

156

absorption, termed heterotriads. The second section presents studies of polynuclear compounds with a surface bound unit that can accept energy from covalently linked chromophoric groups and inject electrons into the semiconductor from its excited state, termed antenna sensitizers.

Heterotriads

A chromophore with a covalently bound electron acceptor or electron donor is commonly referred to as a dyad (*3*). Two simple supramolecular dyad systems, containing a chromophoric component called a sensitizer (S) and a covalently linked acceptor (A) or donor (D) component are shown bound to a semiconductor in Figure 1. When anchored to a semiconductor surface that can act as an electron acceptor, these dyads may be referred to as "heterotriads." In principle, an extension from heterotriads to larger systems can be envisioned keeping in mind that each additional electron transfer step may reduce the Gibbs free energy that can be stored in the final interfacial charge separated state. The heterotriads shown are designed to efficiently perform the steps indicated by 1)

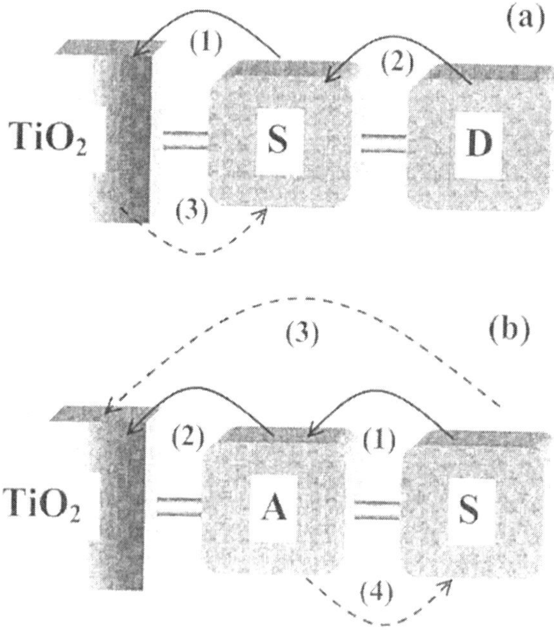

Figure 1. Examples of two "heterotriads" that promote photoinduced intramolecular and interfacial electron transfer. The heterotriads are comprised of TiO₂, a sensitizer, S, and a donor group, D, or acceptor group, A. The desired function of the heterotriads are shown with solid arrows while the potential side reactions are shown with dashed arrows. See the text for more details.

and 2) in Figure 1. The goal is to quantitatively photocreate an interfacial charge-separated pair with an electron in the semiconductor and a "hole" localized on a molecular unit away from the semiconductor surface. The rate constant for electron-hole recombination may be inhibited relative to that observed with a simple molecular sensitizer. The kinetics must be optimized for efficient longlived charge separation to be efficiently achieved. In the heterotriad shown in Fig. la, interfacial excited state electron transfer is expected to be quantitative and ultrafast, so the key to the overall efficiency is likely to be determined by the kinetic competition between the secondary electron transfer step 2) and the primary charge recombination process, step 3). In the second example, on the other hand, the relative kinetics for intramolecular and interfacial electron transfer are expected to control the efficiency, steps 2) and 4) respectively. In both cases, it is evident that considerable control of the factors that govern electron transfer rate constants, such as driving force, reorganizational barriers, and electronic coupling, must be reached before a successful supramolecular device of these types can be developed. The studies described below represent the first steps toward this required molecular precision.

Ru(II) Polypyridine - Phenothiazine Heterotriads

The first dyad synthesized to perform the function shown in Figure la is $Ru(dcb)_2(4-CH_3,4'-CH_2-PTZ,-2,2'-bipyridine^{2+}$, where dcb is $4,4'-(CO_2H)_2-2,2'$-bipyridine and PTZ is the electron donor phenothiazine (*4*). The resulting heterotriad with TiO_2 is shown schematically in Figure 2. Irradiation of the dyad with visible light results in the creation of a metal-to-ligand charge-transfer, MLCT, excited state, which is quenched by electron transfer from the PTZ group in fluid solution. The reductive excited state quenching is moderately exergonic (< 0.25 eV) and has a rate constant of. 2.5×10^8 s^{-1} in methanol, as estimated from the lifetime of the residual *Ru(II) emission. The corresponding charge recombination step is faster than the forward one so that there is no appreciable transient accumulation of the electron transfer product.

When the dyad system is attached to TiO_2, MLCT excitation can result in a new charge separated state with an electron in TiO_2 and an oxidized PTZ group, abbreviated $TiO_2(e^-)$-Ru-PTZ$^+$. In principle there are two possible pathways available to reach this charge separation. In the first pathway, charge injection is followed by oxidation of the phenothiazine donor by the oxidized sensitizer unit, TiO_2-Ru(II)*-PTZ $\rightarrow$ $TiO_2(e^-)$-Ru(III)-PTZ $\rightarrow$ $TiO_2(e^-)$-Ru(II)-PTZ$^+$. In an alternative pathway, reductive quenching by the PTZ group is followed by charge injection into the semiconductor, TiO_2-Ru(II)*-PTZ $\rightarrow$ TiO_2-Ru(I)-PTZ$^+$ $\rightarrow$ $TiO_2(e^-)$-Ru(II)-PTZ$^+$, (*5*). Note that the "Ru(I)" intermediate does not refer to ruthenium in a +1 oxidation state but, rather Ru(II) coordinated to a reduced dcb ligand, i.e. Ru(II)dcb$^-$.

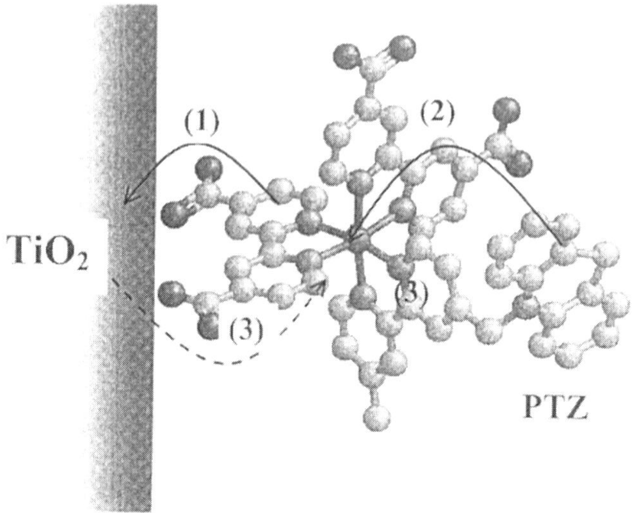

Figure 2. The first heterotriad designed to perform the function shown in Figure 1a, Ru(dcb)$_2$(4-CH$_3$,4'-CH$_2$-PTZ,-2,2'-bipyridine)$^{2+}$, where dcb is 4,4'-(CO$_2$H)$_2$-2,2'-bipyridine and PTZ is the electron donor phenothiazine, abbreviated TiO$_2$-Ru-PTZ. The solid arrows indicate the desired function and the dashed arrow indicates a potentially competitive reaction.

A flash photolysis study of the heterotriad shown in Figure 2 was undertaken (*4*). With nanosecond time resolution it was not possible to determine whether the TiO$_2$(e$^-$)-Ru(II)-PTZ$^+$ state was formed by interfacial electron transfer from the excited or reduced state, i.e. pathway 1 or pathway 2, respectively. However, electron injection into TiO$_2$ from MLCT excited states occurs on a femtosecond time scale, so pathway 1 is the most probable under the experimental conditions employed (*2*). After electron injection, electron transfer from PTZ to the Ru(III) center (-ΔG ~ 0.36 eV) produces the charge separated state TiO$_2$(e$^-$)-Ru(II)-PTZ$^+$. Recombination of the electron in TiO$_2$ with the oxidized PTZ to yield the ground state occurs with a rate constant of 3.6 x 10^3 s^{-1}. Excitation of a model compound that does not contain the PTZ donor, Ru(dmb)(dcb)$_2$$^{2+}$ where dmb is 4,4'-(CH$_3$)$_2$-2,2'-bipyridine, under otherwise identical conditions gave rise to the immediate formation of a charge separated state, TiO$_2$(e$^-$)-Ru(III), whose recombination kinetics were complex and analyzed by a distribution model, with an average rate constant of 3.9 x 106 s^1. Therefore, translating the "hole" from the Ru center to the pendant PTZ moiety inhibits recombination rates by about three orders of magnitude (*4*).

The dyad and model molecules were also tested in regenerative solar cells, with iodide as an electron donor. The photocurrent efficiency was of the order of

45% for both sensitizers at the absorption maximum. However, the open circuit photovoltage, V_{oc}, was observed to be about 100 mV larger for the heterotriad. The effect was even more pronounced in the absence of iodide with 180 mV larger V_{oc} values over 5 decades of irradiance(5). The diode equation, Equation 1, predicts a

$$V_{oc} = \left(\frac{kT}{e}\right) \ln\left(\frac{I_{inj}}{n \sum k_i [A]_i}\right) \qquad (1)$$

59 mV increase in V_{oc} at room temperature for each order of magnitude decrease in the charge recombination rate of injected electrons with acceptors, $k_i[A]_i$, provided that the electron injection flux into the semiconductor, I_{inj} is constant. Applying the spectroscopically measured rate constants to Equation 1 gave a predicted increase in V_{oc} of 200 mV, which was in close agreement with the experimentally determined value of 180 mV. It is remarkable that these molecular interfaces behave like ideal diodes over 5 decades of irradiance with forward electron transfer rates that are at least 6 orders of magnitude faster then charge recombination. Gratzel and coworkers have recently reported an interesting study of heterotriads of this type and have emphasized their potential application in photochromic devices (6). Interestingly, these workers found long-lived charge-separation, like that described for the TiO_2-Ru(II)-PTZ system above, in some cases while not in others. More experiments are required before this interesting interfacial behaviour can be fully understood.

Ru(II)-L-Os(II) Heterotriads

The binuclear compound $[Ru(dcb)_2(Cl)\text{-}BPA\text{-}Os(bpy)_2Cl]^{2+}$, where BPA is 1,2-bis(4-pyridyl)ethane and bpy is 2,2'-bipyridine, shown in Figure 3 was prepared to study processes like that shown in Figure 1a (7). A key difference between this binuclear compound and the Ru(II)-PTZ dyads is that the Os(II) donor is a chromophore that absorbs visible light. The binuclear compound binds to TiO_2 thin films with about 1/2 the surface coverage of a model compound, $[Ru(dcb)_2(Cl)(py)]^+$ where py is pyridine. This suggests that the binuclear complex lies on the nanocrystalline TiO_2 surface in a more or less extended conformation. Transient absorbance difference spectra measured following 532 nm laser excitation, where both Ru(II) and Os(II) chromophores absorb, reveal the typical bleaching of the spin-forbidden MLCT transition localized on the Os(II) group. Spectral and kinetic analysis of the transient signals are consistent with the formation of the charge separated state $TiO_2(e^-)$-Ru(II)-Os(III). This state can be formed through charge injection from the excited Ru chromophore followed by intramolecular Os(II) $\rightarrow$ Ru(III) electron transfer, or via remote electron transfer from the MLCT excited state localized on the $Os^{II}(bpy)_2$ unit. Comparative actinometry revealed that the desired intramolecular electron transfer processes occurs efficiently with 532 or 417 nm

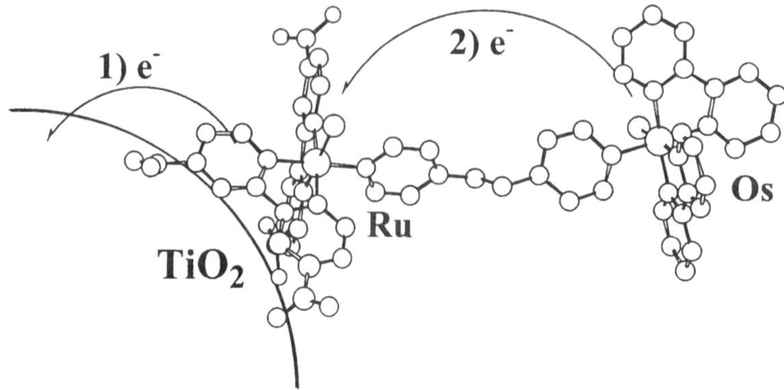

Figure 3. An idealized representation of [Ru(dcb)₂(Cl)-BPA-Os(bpy)₂Cl]²⁺, where BPA is 1,2-bis(4-pyridyl)ethane and bpy is 2,2'-bipyridine, bound to a TiO₂ nanoparticle. The desired electron transfer steps are shown with solid arrows. Surface binding studies suggest that this compound lays flat across the nanoparticle surface and not in the idealized extended geometry shown. See the text for more details.

light excitation. The occurrence of remote transfer from the excited Os unit was confirmed by time resolved experiments after selective laser excitation of the Os(II) chromophore at 683 nm. The lifetime of the $TiO_2(e^-)$-Ru(II)-Os(III) was not significantly different then that of model compounds, presumably because of the semiconductor-dyad orientation.

Precedence for remote electron transfer like that observed in the TiO_2-Ru(II)-L-Os(II) triads exists. In previous work a supramolecular approach for designing a molecular sensitizer with controlled orientation of the component units on the semiconductor surfaces was reported, Figure 4. The binuclear compound is based on a *fac*-Re(I)(dcb)(CO)₃ surface anchoring unit and a -Ru(II)(bpy)₂ chromophore linked through an ambidentate cyanide ligand (*8*). Due to the facial geometry of the surface-bound Re-group, the -Ru(bpy)₂ unit is forced to be proximate to the semiconductor surface.

Visible light excitation of TiO_2 photoanodes, loaded with Re(dcb)(CO)₃(CN)Ru(bpy)₂(CN)⁺ or it linkage isomer, in a regenerative solar cell resulted in efficient light-to-electrical energy conversion (*8*). Plots of the photocurrent efficiency versus excitation wavelength demonstrated that the Ru-polypyridine group absorbs the visible light and converts it efficiently into an electrical current. The transient absorption difference spectra for these sensitizers bound to TiO_2 showed a broad bleach in the region from 400 to 600 nm, expected for the $TiO_2(e^-)$-Re(I)-Ru(III) state. The formation of this state is promptly observed after the laser pulse. This indicates that either remote electron injection into TiO_2, or intraligand (bpy⁻ → dcb) electron hopping from Ru(III)(bpy⁻) to Re(I)(dcb), occurs within the laser pulse ($k > 5$ x 10^8). This data

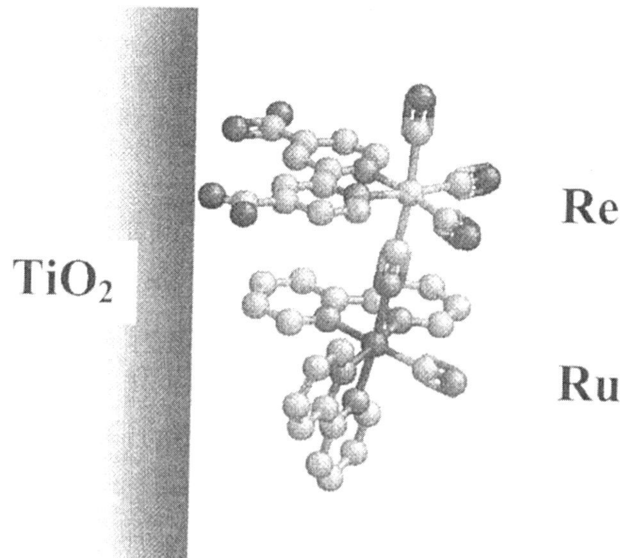

Figure 4. The binuclear compound Re(dcb)(CO)$_3$(CN)Ru(bpy)$_2$(CN)$^+$ attached to a TiO$_2$ surface. The facial geometry about Re(I) orientates the Ru(II) chromophore proximate to the semiconductor surface.

demonstrates a rapid and efficient injection process from a chromophoric group that is not directly coupled to the semiconductor surface. In fact, the solar energy conversion efficiencies realized with TiO$_2$-Re(I)-Ru(II) materials in regenerative photoelectrochemical cells were the same, within experimental error, of that measured for materials sensitized with *cis*-Ru(dcb)$_2$(CN)$_2$. These data indicate that covalent of the sensitizer to the semiconductor surface through chromophoric ligands is not strictly necessary for achieving efficient interfacial electron transfer.

Ru(II)-Rh(III) Polypyridine Supramolecular Compounds

Two novel Ru(II)-Rh(III) polypyridine dyads, containing carboxylic acids groups at the Rh(III) unit, Rh(dcb)$_2$-(BL)-Ru(dmp)$_2$ and Rh(dcb)$_2$-(BL)-Ru(bpy)$_2$ where BL is 1,2-bis[4-(4'-methyl-2,2'-bipyridyl)]ethane) and dmp is 4,7-dimethyl-1,10-phenanthroline, were prepared to function in the manner shown in Figure 1b, (*9*). A dyad and idealized heterotriad are shown in Figure 5. Both in the dyad and heterotriads, the Ru unit was designed to play the role of the chromophoric donor and the Rh unit that of the acceptor. The dyad shown in Figure 5, abbreviated Ru(II)-Rh(III), was studied in considerable detail in fluid solution (*10*). Upon 532.5 nm light excitation of the Ru unit in the dyad, Ru(II)-Rh(III) + hν → *Ru(II)-Rh(III), the typical MLCT emission of this unit is

162

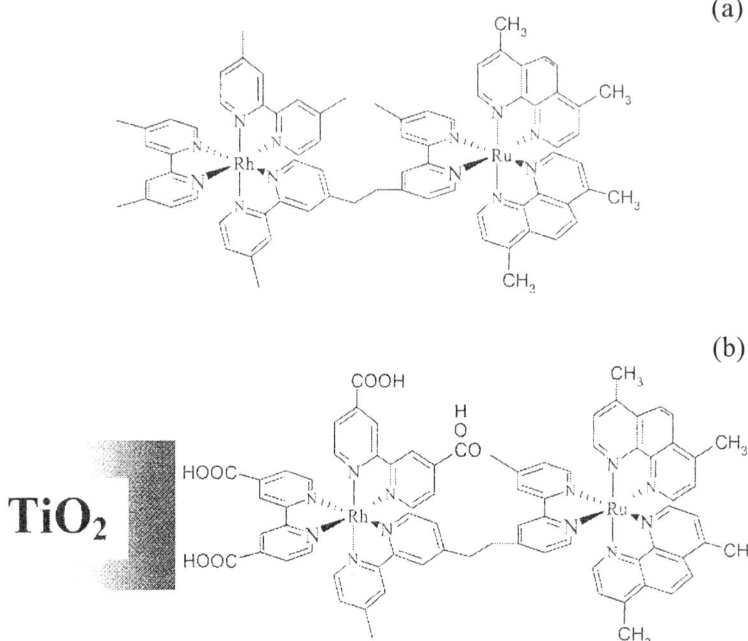

Figure 5. Shown are the dyad Rh(dmb)$_2$-(BL)-Ru(dmp)$_2$ and the heterotriad TiO$_2$-Rh(dcb)$_2$-(BL)-Ru(bpy)$_2$ where BL is 1,2-bis[4-(4'-methyl-2,2'-bipyridyl)]ethane), dmb is 4,4'-(CH$_3$)$_2$-2,2'-bipyridine and dmp is 4,7-(CH$_3$)$_2$-1,10-phenanthroline. The heterotriad was designed to perform the function shown in Figure 1b where the Ru(II) is the sensitizer donor and the Rh(III) group is the acceptor.

strongly quenched with respect to that of a free Ru model, indicating the occurrence of efficient excited-state electron transfer, *Ru(II)-Rh(III) → Ru(III)-Rh(II). An approximate value of the rate constant for this process was estimated from the emission decay, ~ 2 x 10^8 s^{-1} in methanol. Laser flash photolysis at 440 nm, corresponds to excitation of the Ru chromophore, did not show any transient accumulation of electron transfer products. Therefore, the back electron transfer process, Ru(III)-Rh(II) → Ru(II)-Rh(III), is faster than the forward reaction as is often the case for dyads in fluid solution (*3*).

Light excitation at 298 nm is absorbed predominately by the Rh chromophore, Ru(II)-Rh(III) + l*v* → Ru(II)-*Rh(III). Electron transfer from Ru(II) to the excited *Rh(III), occurs with a rate constant, k = 3 x 10^{10} s^{-1}, Ru(II)-*Rh(III) → Ru(III)-Rh(II). The Ru(III)-Rh(II) charge separated state intermediate had a measurable 150 ps lifetime before recombining to yield ground state products.

The kinetics of the three electron transfer steps can be qualitatively rationalized by standard electron transfer theory on the basis of the energetics of the individual components. Summarized in the Jablonski-type diagram in Figure 6 are the relative energetics and lifetimes of the intermediates generated with both 298 and 532.5 nm light, i.e. Rh(III) and Ru(II) excitation respectively. The forward electron transfer following Rh excitation is much faster then that following Ru excitation. Both processes lie in the normal free energy region, but the former is much more exergonic. The fact that the back electron transfer process, which has a very large driving force, appears relatively slow (e.g., slower than the forward process arising from the Rh localized excited state) is most probably a result of Marcus kinetic "inverted" behaviour. The electron transfer kinetics following pulsed light excitation of the Ru chromophore and the covalently bound Rh(III) moiety has also been studied for the Rh'-Ru(dmb)$_2$ and Rh'-Ru(dmp)$_2$ dyads in fluid solution. The observed rate constants for charge separation were: $k = 3.3 \times 10^7$ s^{-1} for Rh'-Ru(dmb)$_2$ and $k = 2 \times 10^8$ s^{-1} for Rh'-Ru(dmp)$_2$. For these dyads, charge separation was slower than the recombination step as shown by a lack of accumulation of the Ru(II)-Rh'(III) electron transfer product.

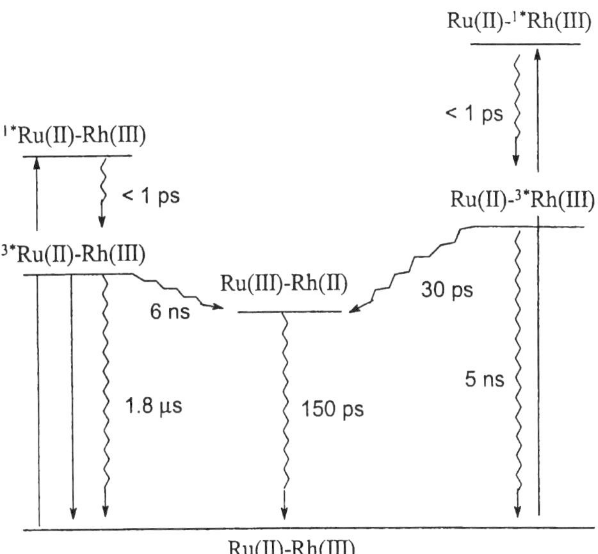

*Figure 6. A Jablonski-type diagram for light excitation of Ru(II) (left hand side) and Rh(III) (right hand side) units in the Rh(III)-Ru(II) dyad. The solid arrows indicate light absorption and the wavy arrows indicate electron transfer or other non-radiative processes. All lifetimes are for the dyad with the exception of the 1.8 µs lifetime of the *Ru MLCT excited state (arrows pointed down on the left hand side), which correspond to a model compound without an appended Rh(III) unit. See discussions in the text for more details.*

The Rh'-Ru dyads anchored to TiO_2 could be used as efficient sensitizers, giving rise to maximum incident photon-to-current efficiency values of ~ 40-50% with iodide electron donors in acetonitrile. The photo-induced electron transfer properties in the heterosupramolecular systems TiO_2-Rh'-Ru(dmb)$_2$ and TiO_2-Rh'-Ru(dmp)$_2$ can be summarized as follows. Upon light excitation of the Ru(II) unit, 1/3 of the surface bound dyads undergo direct electron injection from the excited state of the Ru chromophore, TiO_2-Rh(III)-Ru(II)* $\rightarrow$ TiO_2(e$^-$)-Rh(III)-Ru(III), with a rate constant > 10^8 s^{-1}. The dyads that undergo this remote injection process probably have different surface orientations or accidental contacts in small cavities within the nanocrystalline TiO_2 film. The remaining dyads display stepwise charge injection processes, i.e. TiO_2-Rh(III)-Ru(II)* $\rightarrow$ TiO_2-Rh(II)-Ru(III) $\rightarrow$ TiO_2(e$^-$)-Rh(III)-Ru(III). The first process has comparable rates and efficiencies as for the free dyads in fluid solution. The second step is 40% efficient, because of primary charge recombination between the reduced Rh unit and the oxidized Ru, TiO_2-Rh(II)-Ru(III) $\rightarrow$ TiO_2-Rh(III)-Ru(II). When the recombination of the injected electron and oxidized Ru(III) sites was studied, a remarkable slowing down was observed relative to a simple mononuclear sensitizer, TiO_2-Ru(dcb)$_2$dmb^{2+}. Perhaps more important, these studies demonstrate a 'stepwise' interfacial electron transfer process, like that shown in Figure 1b, for the first time (*9*).

Antenna Sensitizers

The requisites for the supramolecular antenna sensitizers are: (i) an efficient antenna effect, vectorially translating absorbed energy towards a molecular component; and (ii) the capability of the molecular component bound to the semiconductor surface to inject electrons into the semiconductor from its excited state. Antenna sensitizers can increase the fraction of light harvested by a sensitized semiconductor surface. Two simple prototypes, following the "branched" or "one-dimensional" design are shown schematically in Figure 7. An *a priori* evaluation of the two types of designs is difficult, as both have virtues and limitations. For example, when a finite driving force for each energy transfer step is necessary, a branched design, where extensive use of parallel processes is made, is energy saving relative to the one-dimensional design, where all the processes are in series. If, however, fast isoenergetic energy hopping between components can take place, followed by trapping at the lowest-energy component sitting on the surface, then the difference between the two types of design becomes much less critical. In the branched design, the dimensions of the antenna system cannot be increased without introducing energy losses. In the one-dimensional design, quite large antenna systems could be envisioned.

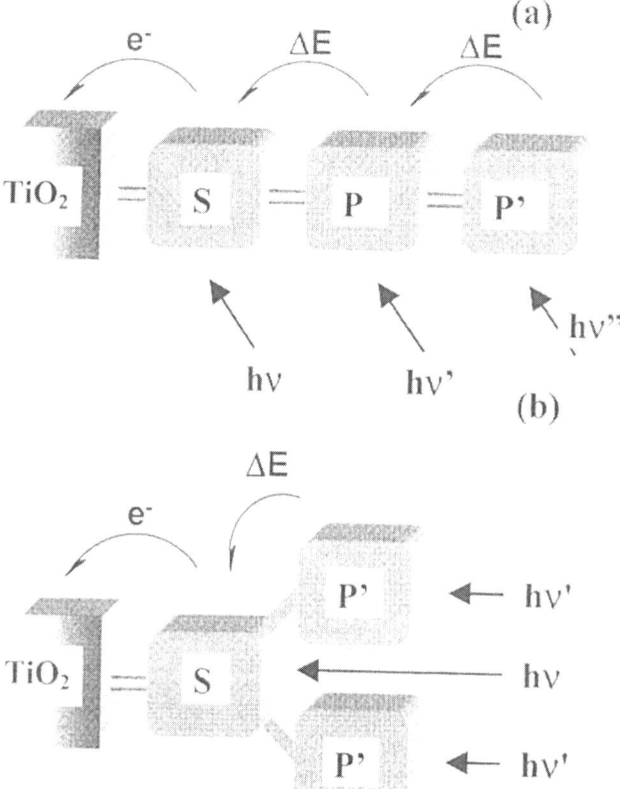

Figure 7. Examples of antenna sensitizers that promote photoinduced intramolecular energy transfer in addition to interfacial electron transfer to TiO₂. The example given in (a) is for a "one dimensional" antenna and that in (b) is for a "branched" antenna.

There are important antenna-semiconductor issues that are specific to the system of interest. For a supramolecular system considered as an independent photochemical molecular device, an obvious expectation is that "the larger the antenna system, the larger the light harvesting efficiency." The same is not necessarily true for light-to-electrical energy conversion on a semiconductor. A highly branched supramolecular system projects a much larger area than a simple molecular sensitizer onto the semiconductor surface. At saturation coverage, this would strongly reduce the potential gain represented by the antenna effect. From this point of view, the one-dimensional design would be superior to the branched one, as one could think of increasing indefinitely the nuclearity of the supramolecular system without substantially increasing the occupied area. These arguments, however, should be taken with caution, as the semiconductor surface available for absorption is far from being an idealized flat surface. In the state-of-the-art nanocrystalline photoanodes, an extremely rough surface is present, with nanometer-sized pores and cavities (2). As such these dimensions may be comparable to those of the antenna systems, geometric considerations may be critical in determining the optimum design type and the degree of nuclearity for a heterosupramolecular antenna sensitizer.

Branched Antenna

The idea of using sensitizer-antenna molecular devices in the sensitization of semiconductors stems from the low efficiency observed with multilayers of sensitizers deposited on electrode surfaces. In a first attempt to investigate the feasibility of such an approach, the [NC-Ru(bpy)$_2$-CN-Ru(4,4'-(CO$_2$H)$_2$-2,2'-bipyridine)$_2$-NC-Ru(bpy)$_2$-CN]$^{2+}$ trinuclear compound was synthesized, Figure 8 (10). Efficient energy funneling from the peripheral chromophores to the central -Ru(4,4'-(CO$_2$H)2-2,2'-bipyridine)$_2$ unit was demonstrated by conventional photophysical experiments and by time-resolved resonance Raman spectroscopy (11,12). Experiments carried out with this branched antennae sensitizer on TiO$_2$ in aqueous solution at pH 3.5, resulted in significant photocurrents. Plots of the photocurrent efficiency versus the excitation wavelength were similar to the absorption spectrum of the trinuclear compound, indicating that the efficiency with which absorbed light was converted to electrons in the external circuit was constant and did not depend on whether the incident light was absorbed by the central unit or by the terminal ones. Subsequent experiments on this complex anchored to nanocrystalline TiO$_2$ gave a global conversion efficiency of ca. 7% under simulated sunlight conditions with turnover numbers of at least five million without decomposition (13). High photocurrent efficiencies were also observed with related compounds based on the same Ru(dcb)$_2$CN)$_2$ core and lateral Ru(l,l0-phenanthroline)antenna (14).

The charge injection process from the photoexcited [NC-Ru(bpy)$_2$-CN-Ru(bpy-4,4'(CO$_2$H)$_2$)$_2$-NC-Ru(bpy)$_2$-CN]$^{2+}$ to the TiO$_2$ semi-conductor was investigated by monitoring the initial decay of the time resolved luminescence (15). From the lifetime of the fast emission component (170 ps), an injection rate of ~ 6 x 10^9 s^{-1} was estimated. As discussed in the previous section, the possibility of remote interfacial electron transfer from the MLCT states localized on the lateral -Ru(bpy)$_2$ units may be an important pathway. Remote electron injection in this case, however would have to compete with the fast intracomponent energy transfer rates known for this class of compounds (16).

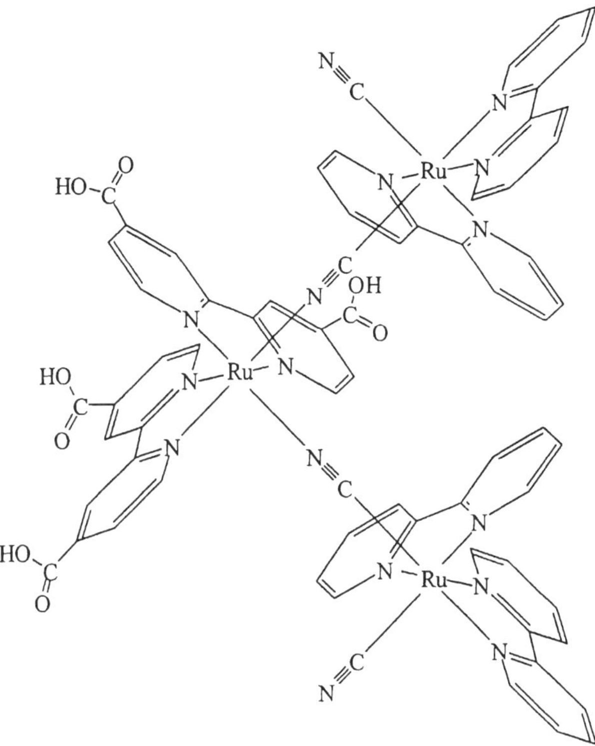

Figure 8. The first branched antenna designed to perform the function shown in Figure 7a, the trinuclear compound [NC-Ru(bpy)₂CN-Ru(4,4'-(CO₂H)₂-2,2'-bipyridine)₂-NC-Ru(bpy)₂-CN]²⁺.

One-Dimensional Antenna

The antenna effect is expected to be of relevance for applications requiring very thin porous films or planar semiconductor surfaces. This is exemplified by a comparison between the photocurrent action spectra of the mononuclear $[Ru(dcbH_2)(CN)_4]^{2-}$ sensitizer and the trinuclear species, $Ru(dcb)(CN)_3(CN)Ru(bpy)_2(CN)Ru(bpy)_2(CN)$ shown in Figure 9. Time resolved resonance Raman and emission spectroscopies show that energy transfer processes, from MLCT excited states localized on the $Ru(bpy)_2(CN)$

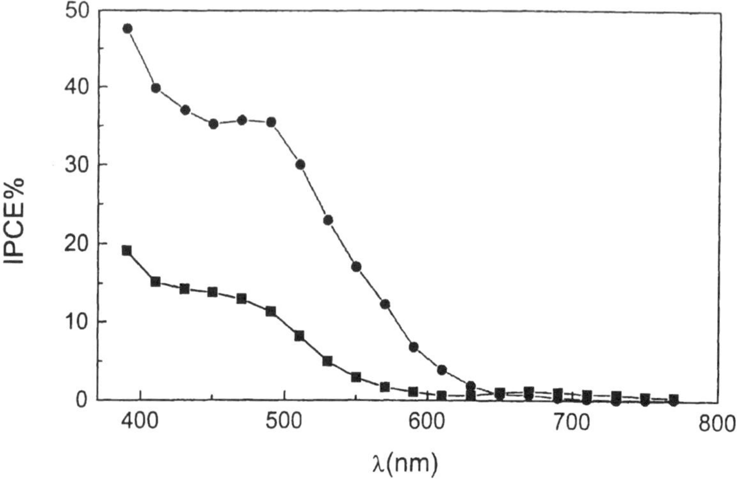

Figure 9. Plots of the incident photon-to-current efficiency percentage, IPCE%, as a function of excitation wavelength (photocurrent action spectra) of the mononuclear [Ru(dcb)(CN)$_4$]$^{2-}$ sensitizer (squares) and the trinuclear species, Ru(dcb)(CN)$_3$(CN)Ru(bpy)$_2$(CN)Ru(bpy)$_2$(CN) (circles) bound to 3 micron thick TiO$_2$ films in regenerative solar cells with 0.5 M LiI/0.05 MI$_2$ acetonitrile electrolyte.

chromophoric units to MLCT states localized on the Ru(dcb)(CN)$_3$ moiety, occur with unit efficiency (*17*). Figure 9 shows the photocurrent efficiency of the mononuclear and one-dimensional trinuclear compounds anchored to ~ 3 µm thick nanocrystalline TiO$_2$ photoanodes in regenerative solar cells with iodide acetonitrile electrolytes. It is evident that the polynuclear sensitizer gives rise to higher photocurrents than the mononuclear species, as expected on the basis of the increased fraction of light that is absorbed. Analogous observations were previously made by comparing the photoaction spectra of Ru(5,5'-dcb)$_2$(CN)$_2$ with the trinuclear Ru(5,5'-dcb)$_2$(CN)(CN)Ru(bpy)$_2$(CN)Ru(bpy)$_2$(CN) sensitizer, where 5,5'-dcb is 5,5'-(CO$_2$H)$_2$-2,2'-bipyridine (*12*). These results confirm that antenna-sensitizer compounds can be profitably used to increase the light harvesting efficiency of sensitized semiconductor materials.

Conclusions

Simple strategies for the design of supramolecular coordination compounds for the conversion of light into an electrical response have been discussed. Artificial heterosupramolecular systems featuring functions such as photoinduced charge separation and the antenna effect have already been utilized in regenerative solar cells and may find applications as components in other molecular photonic devices. The functions that may be realized and their intrinsic efficiency depend critically on the specific choice of molecular components that control the kinetics of interfacial and intercomponent processes. On the basis of the knowledge gained in this field, molecular devices with pre-determined built-in functions will soon be rationally designed and synthesized and integrated into optoelectronic devices.

Acknowledgments

C.A.B. thanks the EU for financial support under the Joule III program, Contract N. JOR3-CT98-7040. G.J.M. gratefully acknowledges The Division of Chemical Sciences, Office of Basic Energy Sciences, Office of Energy Research, U.S. Department of Energy and the National Science Foundation for research support.

Literature Cited

1. a) Qu, P.; Meyer, G.J. in "Electron Transfer in Chemistry", V. Balzani, Ed., John Wiley & Sons; NY **2001**, Chapter 2, Part 2, Vol. IV, pp 355-411. b) Gerischer, H. *Photochem. Photobiol.* **1972**, *16*, 243.

2. Hagfeldt, A.; Grätzel, M. *Chem. Rev.* **1995**, *95*, 49.

3. a) Gust, D.; Moore, T.; Moore, A. L. *Acc. Chem. Res.* **1993**, *26*, 198. b) V. Balzani and F. Scandola, *Supramolecular Photochemistry,* Horwood, Chichester, 1991.

4. (a) Argazzi, R.; Bignozzi, C. A.; Heimer, T. A.; Castellano, F. N.; Meyer, G. J. *J. Am. Chem. Soc.* **1995**, *117,* 11815. (b) Argazzi, R.; Bignozzi, C. A.; Heimer, T. A.; Castellano, F. N.; Meyer, G. J. *J. Phys. Chem. B* **1997**, *101,* 2591.

5. Thompson, D.W.; Kelly, CA.; Farzad, F.; Meyer, G.J. *Langmuir* **1999**, *15,* 650.

6. Bonhote, P.; Moser, J.-E.; Humphry-Baker, R.; Vlachopoulos, N.; Zakeeruddin, S.M.; Walder, L.; Gratzel, M. *J. Am. Chem. Soc.* **1999**, *121,* 1324.

7. Kleverlaan, C.J.; Alebbi, M.; Argazzi, R.; Bignozzi, CA.; Hasselmann, G.M.; Meyer, G.J. *Inorg. Chem.* **2000**, *39*, 1342.

8. Argazzi, R.; Bignozzi, C. A.; Heimer, T. A.; Meyer, G. J. *Inorg. Chem.* **1997**, *36*, 2.

9. Kleverlaan, C. J.; Indelli, M.T.; Bignozzi, C. A.; Pavanin, L.; Scandola, F.; Hasselmann, G. M.; Meyer, G. J. *J. Am. Chem. Soc.* **2000**, *122*, 2840.

10. Indelli, M. T.; Bignozzi, C. A.; Harriman, A.; Schoonover, J. R.; Scandola, F. *J. Am. Chem. Soc.* **1994**, *116*, 3768.

11. (a) Amadelli, R.; Argazzi, R.; Bignozzi, C. A.; Scandola, F. *J. Am. Chem. Soc.* **1990**, *112*, 7099. (b) Smestad, G.; Bignozzi, C. A.; Argazzi, R. *Sol. Energy Mater. Solar Cells* **1994**,*32*, 259.

12. Bignozzi, C. A.; Argazzi, R.; Schoonover, J. R.; Meyer, G. J.; Scandola, F. *Sol. Energy Mater. Solar Cells* **1995**, *38*, 187.

13. O'Regan, B.; Grätzel, M. *Nature* **1991** , *353*, 737.

14. Nazeeruddin, M .K.; Liska, P.; Moser, J.; Vlachopoulos, N.; Grätzel, M. *Helv. Chim. Acta* **1990**, *73*, 1788.

15. Willig, F.; Kietzmann, R.; Schwarzburg, K. *Proc. SPIE Conf. on Energy Efficiency and Sol. Energy Conv. Xl,* 18-22 May, Toulouse, France, SPIE Bellingham, Washington 1992.

16. Schoonover, J. R.; Gordon, K. C.; Argazzi, R.; Woodruff, W. H.; Peterson, K. A.; Bignozzi, C. A.; Dyer, R. B.; Meyer, T. J. *J. Am. Chem. Soc.* **1993**, *115*, 10996.

17. Bignozzi, C. A.; Argazzi, R.; Indelli, M. T.; Scandola, F.; Schoonover, J. R.; Meyer, G. J. *Proc. Indian Acad. Sci. (Chem. Sci.)* **1997**, *109*, 397.

Chapter 13

Ligand Methylation and Coordination Geometry Effects on the Properties of Zinc and Lithium (8-Quinolinolato) Chelate Electroluminescent Materials

Linda S. Sapochak, Flocerfida L. Endrino, Jeffrey B. Marshall,
Daniel P. Fogarty, Nancy M. Washton, and Sanjini Nanayakkara

Department of Chemistry, University of Nevada, 4505 Maryland Parkway,
Las Vegas, NV 89154–4003

We present a study of the photophysical (absorption and emission) and thermal stability properties of zinc bis(8-quinolinolato) (Znq_2), lithium mono(8-quinolinolato) (Liq) chelates, and their methylated derivatives. These materials are compared with aluminum tris(8-quinolinolato) (Alq_3), which has proven to be viable for use in organic light-emitting devices (OLED's). We show that regardless of metal ion substitution the effect of ligand methylation on the photophysical and thermal properties is similar for all Mq_n materials. However, strong solvent dependence of absorption, emission and ^1H NMR chemical shifts for zinc and lithium chelates support a stronger ionic character of the metal-ligand bonding, which may have significant effects on electroluminescent properties.

Introduction

The exploitation of organic electroluminescence (EL) materials as components in organic light emitting devices (OLED's) is a rapidly growing field of study. Since the report by Tang and Van Slyke (*1*) of efficient green EL from aluminum tris(8-quinolinolato) (Alq_3), numerous materials have been studied with emission colors throughout the visible spectrum (*2,3*).

Electroluminescence occurs when electrons and holes are injected and transported through the device where they form excitons near the organic heterojunction, which decay to give light emission. The design of efficient emitter materials requires the optimization of several material properties including: 1) high photoluminescent (PL) efficiency; 2) sublimation without chemical degradation; 3) appropriate energy-level alignment with the cathode and hole-transport layer (HTL) for balanced charge injection and transport; and 4) ability to form structurally stable films. These properties are strongly coupled to the molecular and electronic structure of the molecule, as well as the bulk packing of molecules in vapor deposited films. Alq_3 has a good balance of material properties, which is attributed to its octahedral structure and dipolar nature. Therefore, changes in the coordination geometry and electronic distribution on the ligands should have significant effects on electroluminescent properties of metal (8-quinolinolato) chelates.

We present a study of the photophysical and thermal stability properties of metal (8-quinolinolato) chelates (Mq_n: M = Zn^{+2}, n = 2 and M = Li^{+1}, n=1), and their methylated derivatives, metal (4-methyl-8-quinolinolato) ($4Meq_nM$) and (5-methyl-8-quinolinolato) ($5Meq_nM$). Since the photophysical properties of metal(8-quinolinolato) chelates are centered on the organic ligand (i.e., metal ion is not directly involved in the electron transition leading to PL), we show that the effect of methylation on PL energies and efficiencies is similar to the metal tris(8-quinolinolato) chelates (4). However, unlike Alq_3, the zinc and lithium chelates are shown to exhibit strong solvent dependence of photophysical properties attributed to different coordination geometries and an increased ionic metal-ligand bonding character.

Experimental

Material Synthesis and Characterization

All structures of materials studied are shown in Figure 1.

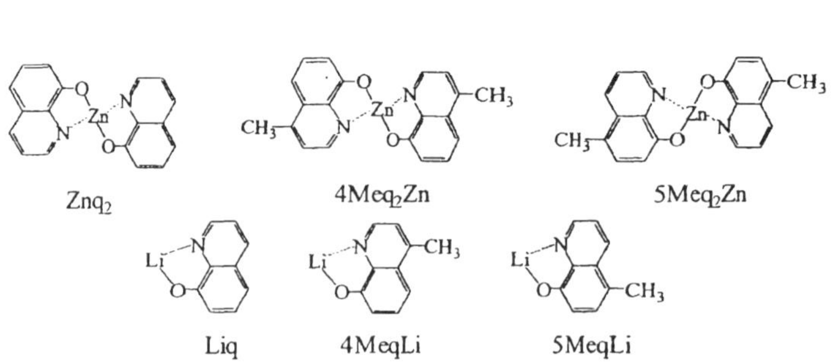

Figure 1. Metal (n-methyl-8-quinolinolato) chelates under study.

Starting materials, 8-quinolinol and reference materials, aluminum tris(8-quinolinolato) (Alq_3) and zinc bis(8-quinolinolato) (Znq_2) were obtained from Aldrich Chemical Co. The 4- and 5-methyl- 8-quinolinol ligands were prepared by modifications of published procedures reported previously (5). ^{1}H NMR was conducted in $CDCl_3$ and DMSO-d_6 using a Bruker 400 MHz NMR. FT-IR spectra were recorded as KBr pellets and as vapor-deposited films on NaCl plates on a Nicolet 210 FT-IR spectrometer. Elemental analysis was obtained from NuMega Resonance Laboratories.

All zinc chelates reported here were prepared by modifications of the synthetic procedure described by Nakamura, et. al., (6) utilizing $Zn(OAc)_2$ dihydrate. The Znq_2 chelates were purified by high-vacuum temperature-gradient sublimation resulting in a < 0.3% deviation from the theoretical % elemental analysis for C, H, and N.

There are few published procedures for the syntheses of lithium mono(8-quinolinolato) chelates (7,8). A general synthetic procedure for preparing Liq and its methyl substituted derivatives from LiOH hydrate is described here. A 10% wt. solution of LiOH hydrate in deionized water was added dropwise to a solution of the appropriate ligand dissolved in minimum amount of absolute ethanol. The pH of the resulting solution was ~8 and precipitate formation occurred almost immediately, with the exception of 4MeqLi. After recrystallization from absolute ethanol, the presence of unreacted ligand was determined by thermal gravimetric analysis (TGA), and was removed by low vacuum sublimation (10^{-3} Torr, ~150-200°C). Further purification was achieved by high-vacuum gradient-temperature sublimation to give a < 0.4% deviation from the theoretical % elemental analysis for C, H, and N.

In contrast to our results, Schmitz, et al (8) reported that adequate elemental analysis could not be achieved for Liq due to its hydroscopic nature. Schmitz prepared Liq from anhydrous LiOH in dry CH_2Cl_2, which resulted in a white crystalline solid that was dried under vacuum but was not further purified. Their TGA analysis showed two weight-loss steps identical to our results for the recrystallized Liq. However, they attributed the first weight loss to dehydration rather than sublimation of unreacted ligand, which is the likely reason for their unreliable elemental analysis.

Zinc bis-(8-quinolinolato) (Znq₂): ^{1}H NMR (ppm, DMSO-d_6) δH2 (s) = 8.47(s); δH3 = 7.34 ; δH4 (d) = 8.27; δH5 = 7.04; δH6 = 7.25; δH7 = 6.87. FT-IR (cm^{-1}, KBr): 1605; 1577; 1500; 1468; 1328; 1242; 1110; 821; 790; 743; and 643. Anal. Calcd for $C_{18}H_{12}O_2N_2Zn$: C, 61.13%; H, 3.42%; N, 7.92%. Found: C–61.39%; H–3.55%; N–7.96%.

Zinc bis(4-methyl-8-quinolinolato) (4Meq₂Zn): FT-IR (cm^{-1}, KBr): 1605; 1578; 1510; 1468; 1316; 1250; 1101; 840; 802; 733; and 669. Anal. Calcd for $C_{20}H_{16}O_2N_2Zn$: C, 62.93%; H, 4.22%; N, 7.34%. Found: C, 62.80%; H, 4.25%; N, 7.32%.

Zinc bis(5-methyl-8-quinolinolato) (5Meq₂Zn): FT-IR (cm^{-1}, KBr): 1602; 1577; 1506; 1466; 1319; 1242; 1093; 833; 786; 755; and 645. $C_{20}H_{16}O_2N_2Zn$: C, 62.93%; H, 4.22%; N, 7.34%. Found: C, 62.69%; H, 4.10%; N, 7.24%.

Lithium (8-quinolinolato) (Liq): FT-IR (cm^{-1}, KBr): 1608; 1573; 1498; 1465; 1321; 1240; 1105; 823; 792; 750; and 648. Anal. Calcd for $C_9H_6O_1N_1Li$: C, 71.54%; H, 4.00%; N, 9.27%. Found: C, 71.29%; H, 4.28%; N, 9.27%.

Lithium (4-methyl-8-quinolinolato) (4MeqLi): FT-IR (cm^{-1}, KBr): 1590; 1572; 1509; 1464; 1326; 1245; 1105; 817; 802; 741; and 668. $C_{10}H_8O_1N_1Li$: C, 72.74%; H, 4.88%; N, 8.48%. Found: C, 72.35%; H, 5.12%; N, 8.50%.

Lithium (5-methyl-8-quinolinolato) (5MeqLi): FT-IR (cm^{-1}, KBr): 1571; 1504; 1467; 1322; 1240; 1089; 832; 784; 757; and 671. Anal. Calcd for $C_{10}H_8O_1N_1Li$: C,72.74%; H, 4.88%; N, 8.48%. Found: C,72.35%; H, 4.70%; N, 8.52%.

Thermal Analysis Characterization

Thermal analysis was determined by differential scanning calorimetry (DSC) and thermal gravimetric analysis (TGA) performed simultaneously using a Netzsch Simultaneous Thermal Analyzer (STA) system. Pure polycrystalline samples (5-8 mg) were placed in aluminum pans and run at a rate of 20°C/min under N_2 gas at a flow rate of 50 ml/min. The temperatures of phase transitions were measured including the melting temperature (T_m). Indium metal was used as the temperature standard. After the first heating, melted samples were cooled at a rate of 20°C/min to form glasses. The glass transition (T_g) and the crystallization point (T_{c1}) were measured from a second heating of the glassy state. The maximum weight loss temperature (decomposition temperature) was determined for each sample run without a lid and reported as the maximum peak of the DTGA (derivative of the TGA curve).

Photophysical Characterization

Absorption and photoluminescence (PL) characterization were performed on dilute (~10^{-5} M) methylene chloride (CH_2Cl_2), dimethyl formamide (DMF), and methanol (MeOH) solutions, as well as vapor-deposited films on quartz slides. Absorption spectra were recorded with a Varian Cary 3Bio UV-vis spectrophotometer and PL spectra were obtained with a SLM 48000 spectrofluorimeter. Extinction coefficients, $\varepsilon(\lambda_{max})$ of the lowest energy absorption peak for Alq$_3$, Znq$_2$, Liq were determined from solutions in CH_2Cl_2 at 10^{-5}-10^{-4} M concentrations where the absorbance at λ_{max} was < 0.5 absorption units. Relative PL quantum efficiencies (ϕ_{PL}) were determined from DMF and CH_2Cl_2 solutions by adjusting the concentration of the sample so that the optical densities at 390 nm (excitation wavelength) were ~0.2. PL quantum efficiencies

(ϕ_{PL}) were calculated relative to the known value for Alq_3 in DMF (ϕ_{PL} = 0.116) (9) and are reported normalized to Alq_3. Samples were run on the same day and all the instrument parameters were kept constant. Each series of samples were run in triplicate.

Results and Discussion

^{1}H NMR Results

The ^{1}H-NMR spectrum of Znq_2 was obtained using a Bruker 400 MHz NMR by heating Znq_2 in DMSO-d_6 directly in the NMR tube to dissolve the material. The resulting NMR spectrum exhibited broad peaks, but the chemical shift values were consistent with those previously reported, by Baker, et al (10). Baker compared the chemical shift values of Znq_2 and Alq_3 and suggested that Zn-O bond may have a stronger ionic character than the Al-O bond based on a larger shielding effect of the aromatic protons for the former.

More recently, Hopkins (11) reported that Znq_2 exhibited two sets of aromatic protons (1:1) in $CDCl_3$ and only one set of broad peaks upon addition of DMSO-d_6. This was explained by suggesting that Znq_2 exists as a tetramer in the solid-state, as reported previously (12). According to single crystal x-ray analysis of a crystal isolated by vacuum sublimation of Znq_2, two crystallographically independent zinc atoms having different coordination geometries were shown to be connected by bridging oxygen atoms to form a tetrameric unit. Hopkins surmised that this tetrameric form of Znq_2 may be stable in the less polar $CDCl_3$ and partially disassociated to the bis-quinolinolato form in DMSO-d_6. Unfortunately, attempts to obtain ^{1}H NMR spectra of Znq_2 in $CDCl_3$ and its methylated derivatives in either $CDCl_3$ or DMSO-d_6 were unsuccessful, due to the insolubility of the materials. We are currently conducting variable temperature ^{1}H NMR experiments in different solvents and powder x-ray diffraction studies to further characterize these materials.

^{1}H NMR studies of the Liq chelates and corresponding ligands were obtained in DMSO-d_6. The proton resonances of all nMeqLi chelates were shielded compared to the neutral ligands, nMeq. The largest shielding effects (~0.4-0.8 ppm upfield shift) occurred for protons H2, H5, and H7 for Liq and 4MeqLi and H2 and H7 for 5MeqLi. These chemical shift values are consistent with those reported by Baker for 8-quinolinol and the 8-quinolinol anion prepared by addition of aqueous NaOH in DMSO (10). These results are reported in Table I.

176

Table I. ^{1}H NMR Chemical Shifts (ppm) for nMeq and nMeqLi in DMSO-d$_6$

Proton	q	Na+/q- Baker(*13*)	Liq	4Meq	4MeqLi	5Meq	5MeqLi
δH2	8.91	8.39	8.45	8.71	8.32	8.87	8.44
δH3	7.59	7.16	7.30	7.41	7.15	7.60	7.34
δH4	8.36	7.92	8.07	----	----	8.40	8.14
δH5	7.45	6.48	6.62	7.48	6.70	----	----
δH6	7.50	7.09	7.20	7.49	7.21	7.28	7.01
δH7	7.19	6.47	6.54	7.10	6.56	6.99	6.35
δCH$_3$	----	----	----	2.66	2.53	2.54	2.38

Table II lists the upfield shift of the aromatic protons of Liq compared to the 8-quinolinol ligand determined in both a polar solvent, DMSO-d$_6$ and a nonpolar solvent, CDCl$_3$. The shielding of the aromatic protons in both solvents suggests that there is delocalization of the lone pair electrons of the phenolic oxygen into the 8-quinolinol ligand upon chelation. This would be true if the Li-O bonding had significant ionic character, where the effect is strongest in the more polar solvent.

Table II. Upfield Shift of Aromatic Protons of q upon Chelation with Li$^+$

Solvent	δH2 (ppm)	δH3 (ppm)	δH4 (ppm)	δH5 (ppm)	δH6 (ppm)	δH7 (ppm)
DMSO-d$_6$	0.46	0.29	0.29	0.83	0.30	0.68
CDCl$_3$	0.32	0.08	0.12	0.78	0.24	0.67

Infrared Spectroscopy Results

The infrared spectra of Znq$_2$, 4Meq$_2$Zn, and Liq have been reported previously (*10,13,14,15*). Consistent with the literature, metal (8-quinolinolato) chelates exhibit similar IR spectra with only small shifts in the peaks due to the metal ion and change in coordination geometry. A comparison of the IR spectra of Znq$_2$ and Liq as KBr pellets and as vapor deposited films are shown in Figure 2. For all metal chelates, no new peaks were observed in the film samples suggesting that no deleterious structural changes occurred upon vapor deposition.

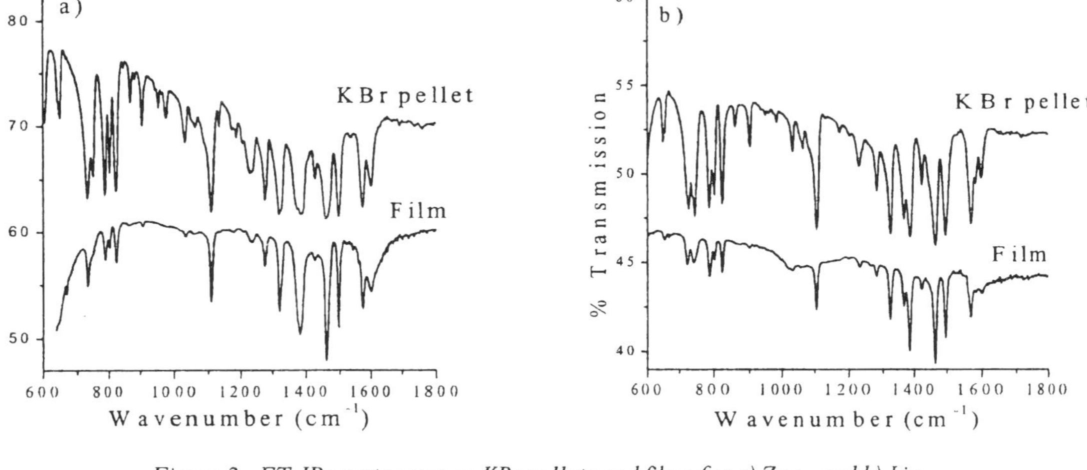

Figure 2. FT-IR spectra ran as KBr pellets and films for a) Znq_2 and b) Liq.

Thermal Analysis Results

The thermal properties of the metal chelates were evaluated by simultaneous thermal analysis (STA), where DSC and TGA are run simultaneously; therefore thermal events observed in DSC can be directly correlated with weight loss events. Results are presented in Table III.

Table III. Thermal Analysis Data

Metal Chelate	T_m (°C)	ΔH_{fusion} (kJ/mol)	T_g (°C)	T_{cl} Onset (°C)	Max. DTGA (°C)
Znq_2	365	26.14	177	272	435
$4Meq_2Zn$	333	24.09	191	n.o.	449
$5Meq_2Zn$	371	24.05	202	324	445
Liq	364	14.32	n.o.	n.o.	518
4MeqLi	397	19.53	n.o.	n.o.	524
5MeqLi	423	21.36	n.o.	n.o.	527

NOTE: n.o. = not observed.

The first heating dynamic of Znq_2 and Alq_3 is shown in Figure 3a, where T_m is offscale so that weaker transitions at lower temperatures can be observed. Znq_2 exhibits two very weak endothermic and one exothermic transition prior to a strong endothermic melting transition (T_m) at 366°C. Similar transitions prior to T_m, but of a stronger nature are observed for metal tris-chelates (Alq_3 and Gaq_3) and were attributed to polymorphic behavior (*4,16*). It is not unlikely that Znq_2 may exhibit polymorphism because at least three polymorphic forms have been observed for Cuq_2 (*17*). Liq, on the other hand, exhibited a single melting transition at 366°C.

Upon methyl-substitution, T_m of the $nMeq_2Zn$ chelates increases in the order $4Meq_2Zn$ < Znq_2 < $5Meq_2Zn$. All zinc chelates exhibited lower melting temperatures than the metal tris-chelates, but the effect of methyl-substitution on T_m were similar. For example, both $4Meq_3Al$ and $4Meq_3Ga$ undergo melting at ~60°C lower temperature than the unsubstituted analogues (*4*), similarly $4Meq_2Zn$ melted at a ~40°C lower temperature compared to Znq_2. In addition, only $5Meq_2Zn$ exhibited an additional transition prior to T_m, suggesting it may also exhibit polymorphism

For $nMeqLi$ chelates T_m increases in the order Liq < 4MeqLi < 5MeqLi. In this series, the 4Meq derivative does not exhibit the lowest melting transition, instead the unsubstituted material, Liq melts ~30°C lower than the methylated derivatives. These trends are more consistent with the melting points of the protonated ligands.

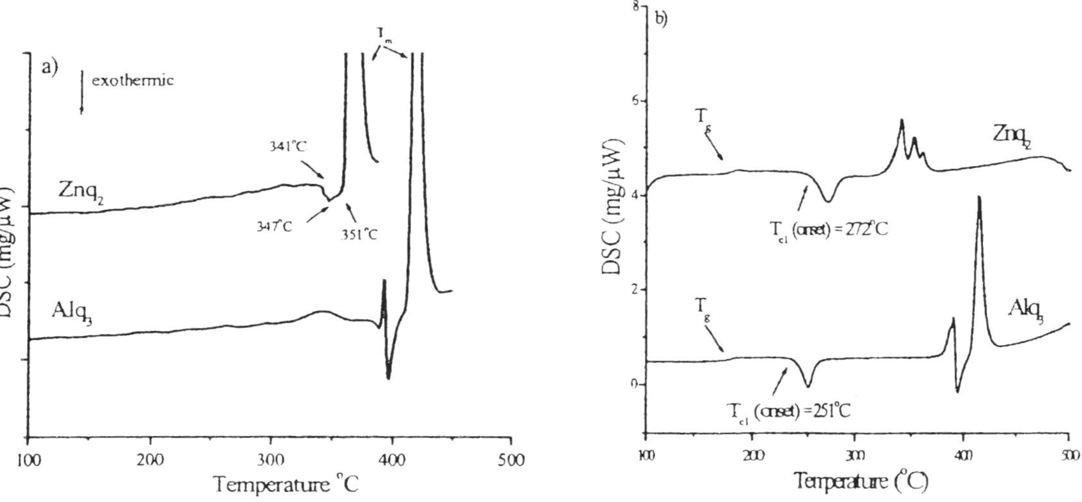

Figure 3. DSC scans for: a) first heating dynamic showing the weaker thermal transitions prior to T_m for Znq$_2$ compared to Alq$_3$; and b) second heating dynamic showing T_g and changes in melting behavior.

180

A second heating dynamic was performed after controlled cooling immediately following T_m. Alq$_3$ and Znq$_2$ revealed glass transitions (T_g) at 177°C and recrystallization exotherms (T_{c1}) at 251°C and 272°C, respectively. The phase changes and melting behavior following T_{c1} for Alq$_3$ were similar to those observed in the first heating dynamic. This was not the case for Znq$_2$, which showed a strengthening of the lower temperature transitions and a weakening of T_m identified in the first heating cycle (see Figure 3b). Whether this thermal behavior is indicative of polymorphic behavior or structural changes of either the bis-quinolinolato chelate or zinc tetramer cannot be determined at this time.

For the methylated derivatives, T_g increases in the order Znq$_2$ < 4Meq$_2$Zn < 5Meq$_2$Zn. The 4Meq$_2$Zn chelate exhibited no recrystallization on heating above the glass transition temperature, whereas 5Meq$_2$Zn showed a broad T_{c1} transition overlapping with a single melting transition following T_g. Interestingly, the glass transition temperatures of Znq$_2$ and the methylated derivatives were almost identical to those of Alq$_3$, and it's methylated derivatives (4). None of the Liq materials formed glasses upon cooling from the melt.

We reported previously, T_m transitions for Alq$_3$ and Gaq$_3$ chelates were accompanied by a ~13-15% weight loss, which was attributed to sublimation of the sample prior to decomposition at temperatures approaching 500°C (4). Znq$_2$ also exhibits a weight loss immediately following T_m, but it was much smaller (~2%) compared to the metal tris-chelates, whereas Liq showed no detectable weight loss. Similar trends were observed when comparing the methyl-substituted derivatives of the metal tris- and bis-chelates, suggesting that, in general the zinc bis(8-quinolinolato) and lithium mono(8-quinolinolato) chelates are more difficult to sublime than the metal tris-chelates.

The thermal stability trends for the nMeq$_2$Zn chelates increases in the order Znq$_2$ < 5Meq$_2$Zn < 4Meq$_2$Zn, and maximum weight loss temperatures (DTGA) were ~50°C lower than the corresponding metal tris-chelates. nMeqLi chelates exhibited the highest thermal stability with maximum of DTGA exceeding 500°C for all materials.

Photophysical Results

Absorption spectra of Alq$_3$, Znq$_2$ and Liq in CH$_2$Cl$_2$ and DMF are depicted in Figure 4.

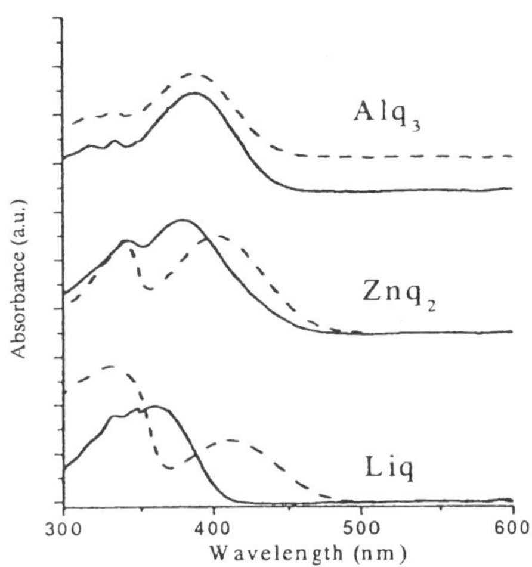

Figure 4. Absorption spectra of Mq$_n$ chelates in CH$_2$Cl$_2$ (solid line) and DMF (dashed line).

Table IV. Absorption Data in Different Solvents

Metal Chelate	CH$_2$Cl$_2$ λ_{max} (nm)	DMF λ_{max} (nm)	MeOH λ_{max} (nm)	Film λ_{max} (nm)	ε λ_{max} in CH$_2$Cl$_2$
Alq$_3$	317, 334, **388**	321, 334, **388**	314, 334[s], **374**	322, 336, **391**	7.86E+03
Znq$_2$	341, **378**	327[s], 339, **402**	321[s], 334, **381**	**342**, 381	4.56E+03
4Meq$_2$Zn	340, **371**	326[s], 339, **398**	321[s], 334, **377**	342, **378**	5.71E+03
5Meq$_2$Zn	346, **393**	336[s], **346**, 419	328[s], 342, **395**	345, 355[s], **397**	4.72E+03
Liq	320,[s] 333,[s] **361**	309[s], **330**, 412	**311**, 374[w]	338, 346, **359**	1.25E+03
4MeqLi	318[s], 334[s], **357**	307[s], **329**, 406	**311**, 372[s]	336[s], 346, **361**	n.d.
5MeqLi	insoluble	309[s], **340**, 428	308[s], **329**, 387[w]	344, **376**	n.d.

NOTE: s = shoulder, w = very weak peak, and n.d. = not determined.

All metal chelates exhibited a broad absorption peak accompanied by one or more higher energy bands in CH_2Cl_2. The shapes of the absorption spectra of Znq_2 and Liq were similar to Alq_3 in CH_2Cl_2, but the longest wavelength transitions (λ_{max}) were blue shifted (682 cm^{-1} and 1928 cm^{-1}, respectively). The extinction coefficients (ε) for λ_{max} were also determined, where ε increased in the order: Liq < Znq_2 < Alq_3. (see Table IV).

There were no differences between the absorption spectra of Alq_3 in CH_2Cl_2 and the more polar solvent, DMF, as reported previously (4). However, both Znq_2 and Liq exhibited large red shifts of the longest wavelength band accompanied by a decrease in intensity relative to the higher energy transitions.

To further investigate the effect of solvent on absorption spectra, samples were run in methanol (MeOH) and as vapor deposited films. Alq_3 exhibited a large blue shift relative to the λ_{max} in CH_2Cl_2 (965 cm^{-1}), whereas both Znq_2 and Liq exhibited red shifts (279 cm^{-1} and 963 cm^{-1}). Absorption spectra of vapor-deposited films were more similar to the spectra in the nonpolar solvent, but absorption peaks were shifted to slightly lower energies probably because of solid-state interactions between the ligands. Results are tabulated in Table IV.

The absorption energy shifts upon methylation of the ligand for $nMeq_2Zn$ and nMeqLi chelates were similar to the metal tris-chelates (4), where the lowest energy absorption band was blue-shifted for C4-methylation and red-shifted for C5-methylation relative to the unsubstituted analogues. The extinction coefficients were determined for the methylated derivatives of Znq_2 and $4Meq_2Zn$ gave the highest value. Solvent dependence of the absorption energies was also observed for all methylated zinc and lithium chelate derivatives (see Table IV).

Table V shows the PL emission energies and relative PL quantum efficiencies (ϕ_{PL}) data for Alq_3, Znq_2 and Liq in CH_2Cl_2 and DMF. The full-width-at-half-maximum (FWHM) of PL emission and the Franck-Condon (F-C) energy shifts (energy difference between absorption and emission) (Δ) are also included. Relative to Alq_3, λ_{max}(emis.) of Znq_2 was red shifted (1681cm^{-1}/DMF and 1240 cm^{-1}/CH_2Cl_2) and Liq was red shifted (869 cm^{-1}) in DMF only. The emission energy of Liq was blue shifted from Alq_3 by 2193 cm^{-1} in CH_2Cl_2. Furthermore, Znq_2, Liq, and their methylated derivatives showed larger F-C shifts in the less polar, and thus less interacting solvent. The F-C shifts for Alq_3 were similar in both solvents. FWHM increased in the order Znq_2 > Alq_3 > Liq, suggesting that Znq_2 is the most vibrationally distorted in its excited state.

Relative PL quantum efficiencies (ϕ_{PL}) increased in the order Liq > Alq_3 > Znq_2 in both solvents. The reported quantum efficiencies of Alq_3 in DMF and CH_3Cl are 0.116 (9) and 0.04 (18), respectively. Unlike Alq_3, both Znq_2 and Liq exhibited higher quantum efficiencies in the less polar solvent. However,

Table V. Photophysical Data in Different Solvents

Metal Chelate	Solvent	Abs. λ(nm)	Emis. λ_{max} (nm)	FWHM (nm)	Δ (cm^{-1})	Rel. ϕ_{PL} (Alq$_3$ =1.0)
Alq$_3$	CH$_2$Cl$_2$	388	514	102	6318	1.0
Znq$_2$	CH$_2$Cl$_2$	378	549	111	8240	0.33
4Meq$_2$Zn	CH$_2$Cl$_2$	371	530	102	8086	1.5
5Meq$_2$Zn	CH$_2$Cl$_2$	393	553	134	7362	0.11
Liq	CH$_2$Cl$_2$	361	491	92	7293	2.6
4MeqLi	CH$_2$Cl$_2$	357	472	82	6825	4.9
5MeqLi	CH$_2$Cl$_2$	---	---	---	---	---
Alq$_3$	DMF	388	521	105	6668	1.0
Znq$_2$	DMF	402	571	110	7362	0.18
4Meq$_2$Zn	DMF	399	546	103	6748	1.2
5Meq$_2$Zn	DMF	419	575	169	6475	0.05
Liq	DMF	410	541	98	5889	1.1
4MeqLi	DMF	405	511	91	5251	3.2
5MeqLi	DMF	419	540	106	5790	0.4

regardless of the metal ion substitution, upon methylation of the 8-quinolinolato ligand ϕ_{PL} increased in the order 4Meq$_n$M >> Mq$_n$ > 5Meq$_n$M, where substitution on the pyridyl ring provided a substantially higher ϕ_{PL} compared to the C5-methylated and unsubstituted analogues. Previously, the increase of ϕ_{PL} for 4Meq$_3$Al vs. Alq$_3$ was attributed to reduced energy loss in the excited state vibrational manifold due to decreased coupling of the metal-ligand stretching coordinates to the electronic transition responsible for PL (*19*). Since the infrared spectra of all metal (8-quinolinolato) chelates are similar this decreased coupling of the vibrational modes to the electronic transition may be a common property for all C4-methylated derivatives.

Solvent Effects on Photophysical Properties

It is reasonable to conclude that in DMF, the red shift of absorption for Liq chelates is caused by disassociation of the chelate by the polar solvent to form a ligand anion (q$^-$) / Li$^+$ pair. Three pieces of experimental evidence support this supposition: 1) the absorption spectrum of deprotonated 8-quinolinol (q$^-$) (DMF/NaOH) was identical to the absorption spectrum of Liq (*20*); 2) the chemical shift values in the ^{1}H NMR spectra of Liq were more shielded in DMSO-d$_6$ vs. CDCl$_3$; and 3) the ϕ_{PL} in CH$_2$Cl$_2$ was larger than in DMF.

Lithium chelates in MeOH showed a dramatic decrease in intensity and red shift of absorption and more than one emission peak in the PL spectra, where the intensities were dependent on the excitation energy. This suggests that more than one emitting species may be present in MeOH solution. For Alq_3 the large blue shift of absorption in MeOH is likely a result of H-bonding with the quinolinolato oxygen, which stabilizes the ground state. The red shift for Znq_2 and Liq chelates in MeOH suggests that H-bonding probably does not occur. Interaction of MeOH with the Zn ion through its oxygen atom, consistent with the formation of the dihydrated form of Znq_2 in water may be likely. It is also possible that a tetrameric structure is stable in MeOH, where the quinolinolato oxygens are blocked and prevented from H-bonding. In either case, the spectral shift observed in MeOH would be more consistent with stabilization of a more polar excited state.

Larger red shifts were observed in DMF for $nMeq_2Zn$ chelates, but it is less likely that Znq_2 disassociates in DMF similar to Liq. Zinc chelates may have a more polar excited state than Alq_3, which is better stabilized in the polar solvent. If the Zn-O bond has more ionic character, the ground state of Znq_2 would be less polar because of delocalization of electron density into the quinolinolato ring by the phenolic oxygen. The fact that Alq_3 shows no significant changes of absorption energy in polar solvents suggests that there is little difference in polarity between the ground and excited states. The possible metal chelate-solvent interactions are depicted in Figure 5, showing that the geometry of the Znq_2 chelates probably changes on interaction with solvent molecules. If the zinc chelates were tetrameric, interaction with DMF could cause dissociation to the monomeric form (Znq_2) as suggested by Hopkins (*11*). Although the present studies do not support nor negate the existence of a tetrameric form, the bis-quinolinolato form is more likely in DMF.

Conclusions

We presented a detailed study of the photophysical and thermal properties of Znq_2, Liq and their methylated derivatives. Similar to Alq_3, C4-methylation of the ligand resulted in substantially higher photoluminescence efficiencies for Znq_2 and Liq. However, Znq_2, Liq, and their methylated derivatives showed strong solvent dependent photophysical properties, which were related to the changes in the coordination geometry about the central metal ion via interactions with solvent molecules, and the more ionic nature of the metal-ligand bonding. Furthermore, the zinc chelates showed similar glass transition temperatures and morphological stabilities compared to Alq_3 and its methylated derivatives, where the Liq materials behaved more like ionic salts. The strong interaction of the Zn

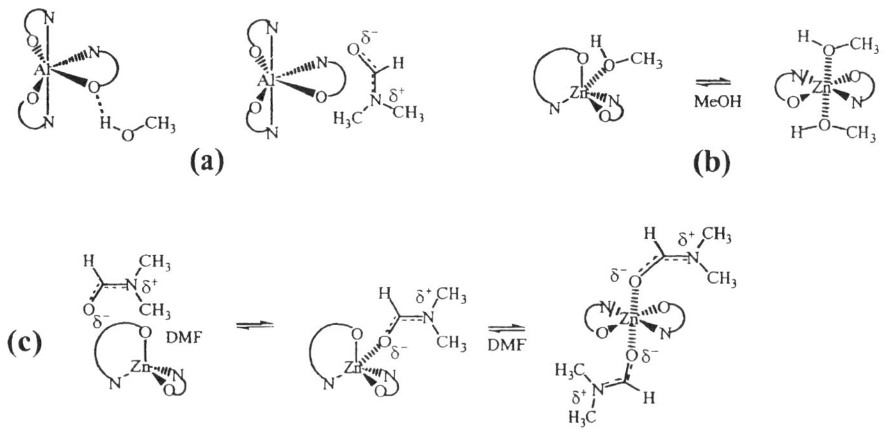

Figure 5. Possible solvent interactions: a) Alq₃ in MeOH via H-bonding with 8-quinolinolato ring oxygen and DMF by dipolar interactions; b) Znq₂ in MeOH via coordination with the Zn ion; and c) Znq₂ in DMF via coordination with the Zn ion coupled with strong dipolar interactions with the ligand.

ion with different solvents suggests that similar interaction of adjacent molecules may be present in the solid state via bridging oxygen atoms (i.e., tetrameric structure). However, better characterization of these materials is necessary before the exact structural nature of Znq_2 in the solid state can be understood. These differences are expected to have a large impact on electroluminescent performance and device studies are currently in progress.

Acknowledgements

The authors gratefully acknowledge NSF (DMR-9874765 grant) and Research Corporation for financial support of this research.

References

1. Tang, C.W. and VanSlyke, S.A. *Appl. Phys. Lett.* **1987**, *51*, 913-915.
2. Chen, C.H. and Shi, J. *Coord. Chem. Rev.* **1998**, *171*, 161-174.

186

3. *Flat Panel Displays: Advanced Organic Materials*; Kelly, S.M. ; Royal Society of Chemistry: Cambridge, UK, 2000; pp. 47-175.

4. Sapochak, L.S.; Padmaperuma, A.B.; Washton, N.M.; Endrino, F.L.; Schmett, G.T.; Marshall, J.B.; Fogarty, D.P.; Burrows, P.E.; and Forrest, S.R. *J. Am. Chem. Soc.* **2001**, *123*, 6300-6307..

5. Padmaperuma, A.B. M.S. of Chemistry, University of Nevada, Las Vegas, NV, 2000.

6. Nakamura, N; Wakabayashi, S; Miyairi, K; Fujii, T. *Chem. Lett.* **1994**, 1741-1742.

7. Shulman, S.G. and Rietta, M.S. *J, Pharm. Sci.* **1971**, *60*, 1762-1763.

8. Schmitz, C.; Schmidt, H. and Thelakkat, M. *Chem. Mater.* **2000**, *12*, 3012-3019.

9. Lytle, F.E.; Story, D.R. and Juricich, M.E. *Spectrochimica Acta.* **1973**, *29A*, 1357-1369.

10. Baker, B. and Sawyer, D. *Anal. Chem.* **1968**, *40*, 1945-1951.

11. Hopkins, T.A.; Meerholz K.; Shasheen, S.;Anderson, M.L.; Schmidt A.; Kippelen, B.; Padias, A.B.; Hall, K.H.; Peyghambrian, N.; Armstrong, N.R. *Chem. Mater.* **1996**, *8*, 344-351.

12. Kai, Y.; Moraita, M; Yasuka, N.; and Kasai, N. *Bull. Chem. Soc. Jpn.* **1985**, *58*, 1631-1635.

13. Tackett, J.E. and Sawyer, D.T. *Inorg. Chem.*, **1964**, *8*, 692-696.

14. Magee, R.J and Gordon Lewis. *Talanta*, **1963**, *10*, 961-966.

15. Charles, Robert.G; Freiser, Henry; Friedel, Robert; Hillyard, Leland E.; and Johnston, William D. *Spectrochimica Acta.* **1956**, 8, 1-8.

16. Brinkmann,M.; Gadret, G.; Muccini, M.; Taliani, C.; Masciocchi, N.; and Sironi, A. *J. Am. Chem. Soc.* **2000**, *122*, 5147-5157.

17. Palenik, G.J. *Acta Crystallogr.* **1964**, *17*, 687-695.

18. Popovych, O. and Rogers, P.G. *Spectrochimica Acta.* **1960**, 49-57.

19. Kushto, G.P.; Iizumi, Y.; Kido, J.; and Kafafi, Z.H. *J. Phys. Chem.* **2000**, *104*, 3670-3680.

20. Endrino, F.L., M.S. of Chemistry, University of Nevada, Las Vegas, NV, 2001.

Chapter 14

Design, Synthesis, and Characterization of Well-Defined Amorphous Molecules for Use in Organic LEDs

Matthew R. Robinson[1,2], Guillermo C. Bazan[2,3,*], Allan J. Heeger[2,4], Marie B. O'Regan[4], and Shujun Wang[3]

[1]Department of Materials Engineering, University of California, Santa Barbara, CA 93106
[2]Institute for Polymers and Organic Solids, University of California, Santa Barbara, CA 93106
[3]Department of Chemistry, University of California, Santa Barbara, CA 93106
[4]UNIAX Corporation, 6780 Cortona Drive, Santa Barbara, CA 93117

Two strategies are presented for making amorphous organic chromophores with well-defined dimensions that exploit the superior qualities of polymers and small molecules with respect to LED fabrication. These qualities are resistance to crystallization, purity, high luminescence efficiency, and high solubility required for spin casting. Tetrakis(4-(4'-(3'',5''-dihexyloxystyryl)styryl)stilbenyl)methane (**T-4R-OC$_6$H$_{13}$**) exemplifies a strategy consisting of four oligophenylenevinylene fragments ("arms") connected to a tetrahedral point of convergence. Bulk samples are amorphous and the film-forming qualities are useful for the fabrication of LEDs with low turn-on voltages. In a related strategy, tris[1-(N-ethylcarbazolyl)-1-(3',5'-hexyloxybenzoyl) methane]-(phenanthroline) europium was designed using a modular approach. It incorporates functionalities for electron and hole transport, solubility, and resistance to crystallization. LEDs were fabricated and studied.

Tetrahedral Oligophenylenevinylenes

There is considerable research activity centered on organic chromophores with geometric attributes that discourage crystallization. Examples include conjugated polymers,(1) "starburst", "spiro", binapthyl, and tetrahedral structures.(2,3,4) Work from our labs has shown that attaching four optoelectronic organic fragments to a tetrahedral point of convergence often results in an amorphous material (5) and that these can be used in the fabrication of light emitting diodes with low turn-on voltages (6).

crystalline

amorphous

Molecules with more extended conjugation are also amorphous. For example, we recently reported on the morphology and low voltage electroluminescence of tetrakis(4-(4'-(3'',5''-dihexyloxystyryl)styryl)stilbenyl)methane (**T-4R-OC$_6$H$_{13}$**), shown below.(6)

T-4R-C$_6$H$_{13}$

While a disordered (amorphous) arrangement is required to generate good quality thin films for incorporation into devices, it is not always a sufficient condition. Additionally, within a thin film, the material can crystallize at different rates, or to a greater extent than in the powder form.(7) The solvent from which the film is cast becomes an important parameter (8) because it may facilitate motion within the material or may form part of a crystalline lattice not readily achieved by the neat material. Therefore, to evaluate the potential of organic glasses in device configurations one must examine the film properties and the stability toward crystallization as a function of molecular structure and sample history.

With this in mind, the molecules tetrakis(4-(4'-(2'',5''-dioctyloxy-4''-(4'''-(2'''',5''''-dioctyloxy-4'''''-styryl)styryl)styryl) styryl)styryl) phenylmethane (**T-6R-OC₈H₁₇**) and 1-[(4'-(2'',5''-dioctyloxy-4''-(4'''-(2'''',5''''-dioctyloxy-4'''''-styryl)styryl)styryl)styryl)styryl]benzene (**6R-OC₈H₁₇**) were prepared to probe the effect of a tetrahedral molecular shape on bulk morphology.(9) Recent studies have also revealed the importance between topology and the nature of interchain or interchromophore interactions in polymers and small molecules. Fluorescence microscopy and AFM probe the topology and prove informative for understanding the morphology of the thin film and how intermolecular packing arrangements affects optical properties.(8,10,11) Note that **T-6R-OC₈H₁₇** has a mass of 4415 amu. It is thus a well defined macromolecule of intermediate dimensions.

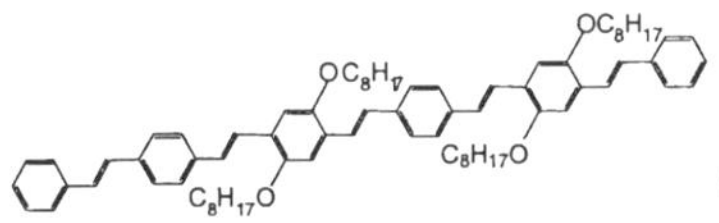

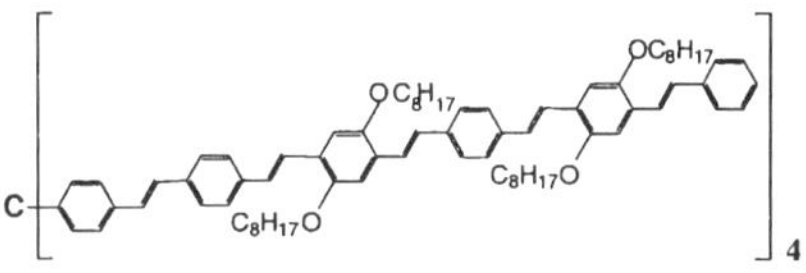

No evidence of crystallinity in bulk samples of **T-6R-OC₆H₁₃** could be detected by X-ray powder diffraction. DSC analysis shows a weak melting transition at 191 °C. The magnitude of the transition is consistent with packing between the octyloxy substituents, rather than long range order of the larger tetrahedral molecules. AFM studies show that films of **T-6R-OC₈H₁₇** obtained by casting a chloroform solution show regular surface patterns. By heating to 190 °C one obtains smooth homogenous films. In contrast, heating a thin film of **6R-OC₈H₁₇** results in the formation of macroscopic crystals Therefore by incorporating the optoelectronic unit into a tetrahedral array, one can change the bulk properties from crystalline to amorphous.

The photoluminescence (PL) quantum yield of **T-4R-OC₆H₁₃**, in chloroform (ϕ_{sol}) and as a thin film (ϕ_{neat}) are 0.67 and 0.42 respectively. The combination of high solid state quantum yield and good film forming tendencies prompted us to incorporate **T-4R-OC₆H₁₃** within a device structure. Stable electroluminescence is observed at 3 V, without the use of an anode buffer, however the best efficiencies are achieved with the configuration ITO/PEDT/PVK/**T-4R-OC₆H₁₃**/Ba/Al. With this configuration electroluminescece turns on at 3 V, and exhibits a brightness of 558 cd/m^2 at 5.4

V (22 V/μm) with an external quantum efficiency of 0.01 photons per electron. No attempt towards optimization has been made however, and higher efficiencies are anticipated after balancing charge injection by using electrode materials that better match the HOMO and LUMO energies of the electroluminescent layer.

LED performance was also studied as a function of film quality using **T-6R-OC$_8$H$_{17}$**. Heating likely perturbs the ITO-organic interface, so no correlation exists between device performance and the surface quality observed by AFM.(9)

Modular Molecular Design of a Europium-Based LED

One disadvantage of organic LEDs is their characteristically broad emission profile. Sharp, emission bands are desirable for use in full color displays. Pure red emission is characteristic of Europium tris(β-diketonate) complexes, since the emission from these complexes stems from relaxation between f levels that are well protected from environmental perturbations by filled $5s^2$ and $5d^{10}$ orbitals.(12)

Most europium-based complexes readily crystalize, have low carrier mobility, and exhibit self-quenching, making them less than ideal candidates for devices. Lanthanide-containing LEDs have been fabricated by vacuum deposition of the complex onto a suitable charge injection layer and the best devices take advantage of a multilayer architecture that optimizes charge balance.(13) The combination of conjugated polymers doped with europium complexes has also recently appeared.(14) An alternative approach makes use of a modular approach at the molecular level for ligand design. Tris[1-(N-ethylcarbazolyl)-1-(3',5'-hexyloxybenzoyl)methane] [phenanthroline]europium (**Eu(M$_1$)$_3$Phen**) incorporates hole and electron transporting moieties, as well as side groups for solubility and morphological control.(15)

Eu(M$_1$)$_3$Phen

In this molecule, the alkoxy chain and overall molecular shape combine to give stable glasses directly from solution. **Eu(M₁)₃Phen** shows a glass transition temperature at 65°C by DSC. No evidence of crystallinity is observed with X-ray powder diffraction. It is readily souble in toluene and clear transparent films are obtained by spin casting from solution. The phenanthroline ligand was included since it is thought to be an electron transport functionality(16), while the carbazole fragment was incorporated to improve hole transport.(17)

The absorption spectrum of **Eu(M₁)₃Phen** shows an onset at 460 nm with a peak at 380 nm, identical to that of the protonated ligand. The emission maximum occurs at 612 nm, and is characteristic of the europium ion within a tris(β-diketonate)/phenanthroline ligand environment. There is no residual emission from the ligand and the photoluminescence quantum yield is approximately 0.50, as measurend by a calibrated integrating sphere.

An LED with the architecture ITO/**Eu(M₁)₃Phen**(60 nm)/Ca was made by spinning **Eu(M₁)₃Phen** onto ITO directly from toluene, followed by thermal evaporation of calcium at 10^{-6} Torr. There were no attemps towards optimization so the best test for a successful design is to compare the performance to other successfully-used europium chromophores such as tris(dinaphthoylmethane)-(phenanthroline) europium (**Eu(2-2)₃Phen**).

Eu(2-2)₃Phen

Figure 1 (top) shows the light intensity-current-voltage (LIV) curve and device configuration of the single component LED. Also shown is the efficiency (ϕ). Note the low turn-on voltage (5 V) of the **Eu(M₁)₃Phen** device. Furthermore, note that the device made with **Eu(2-2)₃Phen** failed e.g. no light output was observed. In order to compare the efficiency of **Eu(M₁)₃Phen** to that of **Eu(2-2)₃Phen**, it was necessary to use a PVK layer. *Figure 1 (bottom)* shows the LIV data for LEDs with the configuration ITO/PVK (30 nm)/**Eu(M₁)₃Phen**(60 nm)/Ca. The efficiency of **Eu(M₁)₃Phen** is over ten-fold higher than that of **Eu(2-2)₃Phen**. Furthermore, the emission from the **Eu(M₁)₃Phen** device is stable for hours under a nitrogen atmosphere, while for **Eu(2-2)₃Phen**, the devices last less than a few seconds. We suspect that the **Eu(M₁)₃Phen** device lasts longer when PVK is included because of the low glass transition temperature (T_g) associated with the neat compound. Through modular design, the T_g can be increased, for example by shortening the aloxy side groups.

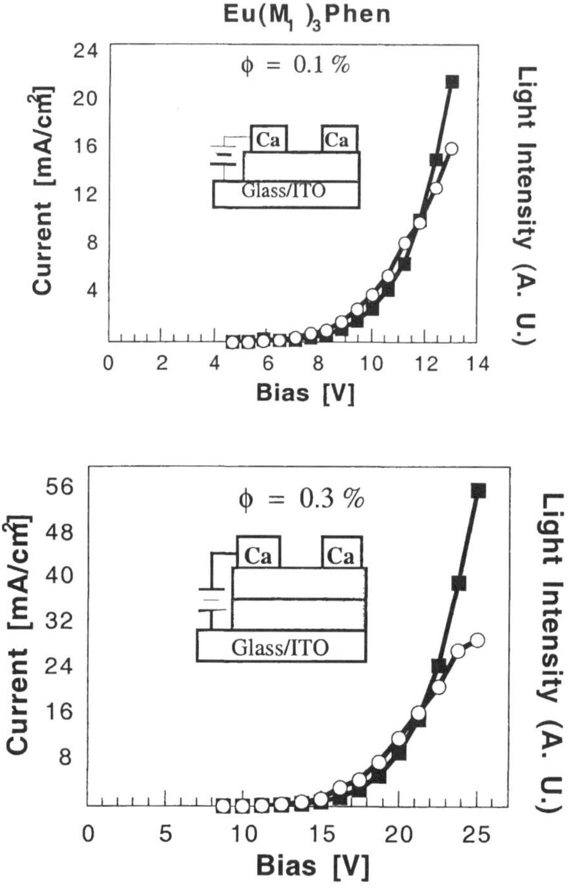

*Figure 1. Voltage versus current(■) and light intensity (○, right axis) for **Eu(M₁)₃Phen** with the shown device archetecture. Also shown is the device efficiency for both **Eu(M₁)₃Phen** and **Eu(2-2)₃Phen**.*

In summary, we have shown that it is possible to introduce geometrical attributes to molecules of intermediate dimensions which render the bulk material resistant to crystallization. A comparison of **T-4R-OC$_6$H$_{13}$** against **6R-OC$_8$H$_{17}$** demonstrates one strategy. The tetrahedral shape of **T-4R-OC$_6$H$_{13}$** makes the molecule awkward to pack and the increased molecular shape retards molecular motions which are necessary to find a lattice arrangement. In the case of **Eu(M$_1$)$_3$Phen**, the overall complex shape, in cooperation with the aliphatic chains in the ligand provide an amorphous solid. This europium complex incorporates "optoelectronic functionalities" within the ligand framework, which enables fabrication of a single layer LED.

References

1. Kraft, A.; Grimsdale, A. C.; Holmes, A. B. *Angew. Chem., Int. Ed. Engl.* **1998**, *37*, 402.
2. (a) Shirota, Y.; Kobata, T.; Noma, N. *Chem. Lett.* **1989**, 1145. (b) Inada, H.; Shirota, Y. *J. Mater. Chem.* **1993**, *3*, 319. (c) Ueta, E.; Nakano, H.; Shirota, Y. *Chem. Lett.* **1994**, 2397. (d) Inada, H.; Ohnishi, K.; Nomura, S.; Higuchi, A.; Nakano, H.; Shirota, Y. *J. Mater. Chem.* **1994**, *4*, 171. (e) Kageyama, H.; Itano, K.; Ishikawa, W.; Shirota, Y. *J. Mater. Chem.* **1996**, *6*, 675. (f) Tanaka, H.; Tokito, S.; Taga, Y.; Okada, A. *Chem. Commun.* **1996**, 2175. (g) Tanaka, S.; Iso, T.; Doke, Y. *Chem. Commun.* **1997**, 2063. (h) Naito, K.; Miura, A. *J. Phys. Chem.* **1993**, *97*, 6240. (i) Naito, K. *Chem. Mater.* **1994**, *6*, 2343. (j) Naito, K.; Sakurai, M.; Egusa, S. *J. Phys. Chem. A* **1997**, *101*, 2350. (k) Kraft, A. *Chem. Commun.* **1996**, 77. (l) Bettenhausen, J.; Strohriegl, P. *Adv. Mater.* **1996**, *8*, 507. (m) Bettenhausen, J.; Greczmiel, M.; Jandke, M.; Strohriegl, P. *Synth. Met.* **1997**, *91*, 223. (n) Salbeck, J. *Ber. Bunsenges. Phys. Chem.* **1996**, *100*, 1667. (o) Salbeck, J.; Yu, N.; Bauer, J.; Weissörtel, F.; Bestgen, H. *Synth. Met.* **1997**, *91*, 209. (p) Wang, P.-W.; Liu, Y.-J.; Devadoss, C.; Bharathi, P.; Moore, J. S. *Adv. Mater.* **1996**, *8*, 237. (q) Meier, H.; Lehmann, M. *Angew. Chem. Int. Ed. Engl.* **1998**, *37*, 643. (r) O'Brian, D. F.; Burrows, P. E.; Forrest, S. R.; Koene, B. E.; Loy, D. E.; Thompson, M. E. *Adv. Mater.* **1998**, *10*, 1108. (s) Koene, B. E.; Loy, D. E.; Thompson, M. E. *Chem. Mater.* **1998**, *10*, 2235. (t) Hu, N.-X.; Xie, S.; Popovic, Z.; Ong, B.; Hor, A.-M. *J. Am. Chem. Soc.* **1999**, *121*, 5097.
3. Johansson,N.; Salbeck, J.; Bauerc, J.; Weissortel, F.; Broms, P.; Andersson, A.; Salaneck, W. R. *Adv. Mater.* **1998**, *10*, 1137.
4. Wang,S.; Oldham, W. J. Jr.,; Hudack, R. A. Jr.; Bazan, G. C. *J. Am. Chem. Soc.* **2000**, *122*, 5695.

5. For related work on polymeric materials, see: Sunder, A.; Heinemann, J.; Frey, H. *Chem. Eur. J.* **2000**, *6*, 2499.

6. Robinson,M. R; Wang, S.; Bazan, G. C. *Adv. Mater.* **2000**, *12*, 1701.

7. Sauer,B. B.; McLean, R. S.; Thomas, R. R. *Polymer International* **2000**, *49*, 449.

8. Conboy,J. C.; Olson, E. J. C.; Adams, D. M.; Kerimo, J.; Zaban, A.; Gregg, B. A.; Barbara, P. F. *J. Phys. Chem. B* **1998**, *102*, 4516.

9 Robinson, M.R.; Wang, S.; Heeger, A. J. Bazan, G.C. *Advanced Functional Mateials*, submitted..

10. (a) Dittmer,J. J.; Lazzaroni, R.; Leclère, P.; Moretti, P.; Granström, M.; Petritsch, K.; Marseglia, E. A.; Friend, R. H.; Brédas, J. L.; Rost, H.; Holmes, A. B. *Solar Energy Materials & Solar Cells* **2000**, *61*, 53. (b) Lee, T-W.; Park, O. O. *Adv. Mater.* **2000**, *12*, 801.

11. Han,E.-M.; Do, L.-M.; Fujihira, M.; Inada, H.; Shirota, Y. *J. Appl. Phys.* **1996**, *80*, 3297.

12. (a) Sinha, A. P. B., in *Spectroscopy in Inorganic Systems*, Vol. 2; Rao, C. N. R.; Ferraro, J. R. Eds.; Academic Pres, New York, 1971. (b) Reisfeld, R.; Jorgensen, C. K. *Lasers and Excited States of Rare Earths*, North Holland, New York, 1978.

13. Campos, R. A.; Kovalev, I. P.; Guo, Y.; Wakili, N.; Skotheim, T. *J. Appl. Phys.* **1996**, *80*, 7144.

14. McGehee, M. D.; Bergstedt, T.; Zhang, C.; Saab, A. P.; O'Regan, M. B.; Bazan, G. C.; Srdanov, V. I.; Heeger, A. J. *Adv. Mater.* **1999**, *11*, 1349.

15. Robinson, M.R.; O'Regan, M.B.; Bazan, G.C. *Chem. Commun.* **2000**, 1645.

16. Kido, J.; Ineda, W.; Kimura, M.; Nagai, K. *Jpn. J. Appl. Phys.* **1996**, *35*, L 394.

17. Liu, L.; Li, W.; Hong, Z.; Peng, J.; Liu, X.; Liang, C.; Liu, Z.; Yu, J.; Zhao, D. *Synth. Metals* **1997**, *91*, 267.

Theory

Electronic Coupling of Donor–Acceptor Sites Mediated by Homologous Unsaturated Organic Bridges

Marshall D. Newton

Chemistry Department, Brookhaven National Laboratory, Upton, NY 11973–5000

Electronic coupling matrix elements (H_{DA}) elements for a number of radical cation and anion systems of type $DBA^{\pm}$, where D,B, and A denote donor, bridge and acceptor moieties, have been evaluated and analyzed, using configuration interaction electronic structure calculations (INDO/s) in conjunction with the Generalized Mulliken Hush model, which also yields effective D/A separation distances (r_{DA}). The systems studied include D/A = CH_2 or ferrocenyl, and B = oligovinylene (OV), oligo-p-phenylene-ethynylene (OPE), oligo-p-phenylene-vinylene (OPV), or oligomethylene (OM). Coupling based on alternative hole-states is considered and electronic structural features relevant to D/A coupling are compared for ferrocene and a cobalt-based analog.

Long range donor(D)/acceptor(A) interactions play a crucial role in controlling kinetic and spectroscopic properties of many molecular systems (1-4). For example, the rate of intramolecular electron (or hole) and energy transfer between localized D and A sites linked by a molecular bridge (B) may be proportional to the square of the effective D/A electronic matrix element (H_{DA}) in cases of relatively weak coupling, as is frequently the case in long-range transfer between sites separated by distances large relative to the size of the D and A sites. This situation, corresponding to coherent electronic tunnelling (1,3,5), represents a special limiting case among a broad range of charge and energy transport mechanisms, which is controlled by a delicate balance of factors, including (1) the magnitude of the energy gaps separating the initial and final states from intermediate states, in which the transferring charge or excitation resides on B; (2) the strength of coupling between the D,A groups and the bridge (B); and (3) relaxation dynamics involving the nuclear modes of DBA system and the surrounding medium. Figure 1 illustrates schematically the case for thermal charge transfer of the charge separation type,

$$DBA \rightarrow D^{\pm}BA^{\mp} \tag{1}$$

(where the D and A levels are resonant in the transition state) in the case of a 2-site bridge, distinguishing "electron" (*et*) and hole (*ht*) transfer processes, which occur in parallel and are subject to quantum interference (either constructive or destructive). Except as noted below, *et* is used collectively to refer to *et* and *ht* processes. As the gaps to the intermediate bridge states decrease, one expects (5) a transition from coherent tunnelling via superexchange with only "virtual" residence of charge on the bridge, to incoherent sequential transfer involving finite lifetimes for the intermediate states: i.e., $D^{+}B^{-}A$ (*et*) and $DB^{+}A^{-}$ (*ht*), in the example given in Figure 1. Within the bridge, the intermediate states may be delocalized or localized on particular bridge sites (two sites are represented in Figure 1). In the limit of coherent tunneling, simple theoretical models (e.g., based on tunneling through a 1-dimensional rectangular barrier or on superexchange theory) predict an exponential falloff of H_{DA} with effective D/A separation, r_{DA} (4):

$$H_{DA} \propto \exp\left(\beta r_{DA} / 2\right) \tag{2}$$

(the factor of ½ in the exponent reflects the convention that β governs the decay of H_{DA}^2, in view of the presence of the latter in the rate constant prefactor for weak-coupling (Golden Rule) kinetics). Of course, with D and A groups of finite extent there is in general no unique definition of r_{DA}, a topic to which we return below.

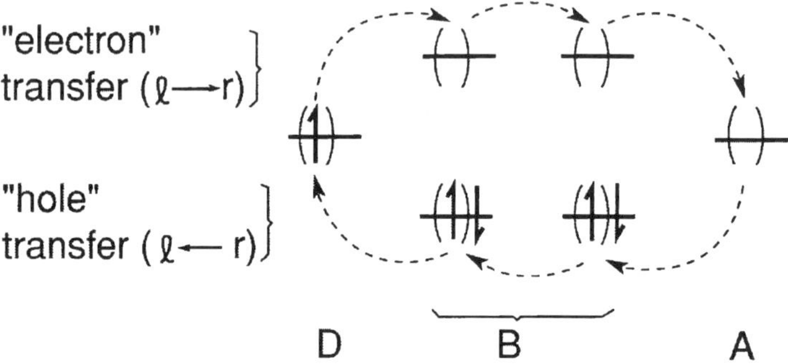

Figure 1. Schematic representations of thermal bridge-mediated charge transport of the "electron" (top, left to right) and "hole" (bottom, right to left). type, illustrated for the case of two intervening bridge units. The intermediate states for the hole and electron processes involve, respectively, charge-localization based on the filled ("valence") and empty ("conduction") bands of the bridge. When the energy gaps separating the common (resonant) donor/acceptor (D/A) level from those of the bridge (taken here as equal, as in a homologous spacer) are large relative to the magnitude of coupling between D and A groups and the bridge, coherent (superexchange) tunnelling occurs between D and A, and the intermediate bridge states are "virtual". As the gaps become smaller, the residence lifetimes of the excess charge on the bridge sites increases, leading eventually to a conventional sequential hopping process from D to A.

The main objective of this paper is to evaluate theoretically H_{DA} magnitudes for *et* mediated by a variety of electronically unsaturated oligomeric organic bridges. On the basis of the calculated results, we assess the dependence of

H_{DA}, and especially its rate of decay with bridge length (see eq 2), on the nature of the D,A, and B groups, including the influence of conformational fluctuations and substituent effects at bridge sites. Furthermore, we consider cases of near degeneracy (as, for example, when transition metal complexes serve as D and A groups), in which more than one pair of initial and final charge transfer states, with correspondingly distinct H_{DA} magnitudes, may be accessible. The particular DBA systems selected for study (depicted in Figure 2) are for the most part model systems related to those for which interfacial electrochemical *et* kinetics have been measured at self-assembled-monolayer(SAM)-film-modified gold electrodes (6-9).

Before proceeding to deal with specific *et* processes, we note that to first approximation, the matrix elements (H_{DA}) governing *et* or *ht* (and related conductive behavior) may be taken as *effective* 1-particle (orbital) matrix elements (4). This is in contrast to other types of D/A interactions, such as energy transfer, and magnetic exchange, in which H_{DA} is a 2-particle matrix element (in "double" exchange, both 1- and 2- particle interactions are involved).

The magnitude of the "coupling element" is important in determining (10) whether a given charge transfer mechanism is non-adiabatic (governed by the Golden Rule) or adiabatic (pertinent for strong coupling). However, it is important to understand the distinction between the overall matrix element (generally vibronic) relevant in this sense, and the purely electronic coupling element (H_{DA}). As noted above, H_{DA} may be viewed, at least approximately, as a 1-particle (orbital) matrix element. However, many-electronic quantum chemical calculations (of the type employed here) have shown that the spatial H_{DA} may be displayed as a dominant orbital factor scaled by an "electronic" Franck Condon factor, typically ~0.9, which reflects many-electron screening which accompanies the 1-particle transfer (2,11). In addition, the spatial H_{DA} may be scaled by a factor depending on the overall spin state of the DBA system (4), an issue of particular interest in the case of open-shell transition metal complexes. Finally, in the overall vibronic coupling element $(H_{DA})_{eff}$, the electronic factor (H_{DA}) is scaled by a nuclear Franck-Condon factor (10). In the present paper we deal with the spatial H_{DA} elements, which implicitly include any many-electron effects entailed in the quantum chemical calculations.

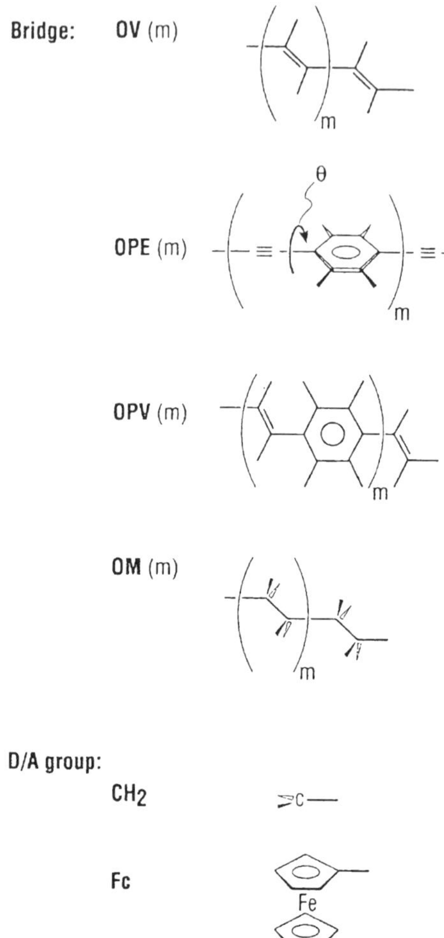

Figure 2. Molecular structures of the DBA type for which calculations were carried out. Three types of unsaturated bridges were employed: oligovinylene (OV), with a vinylene unit plus m V repeat units (m = 1,3,5); oligo-p-phenylenethynylenes (OPE), with an ethlynylene (acetylene) unit plus m PE repeat units (m = 0-4); and oligo-p-phenylenevinylene (OPV), with a vinylene unit plus m PV repeat units (m=0-4). The saturated oligomethylene (OM) bridges (in fully staggered conformation) are based on a (CH₂)₂ unit plus m additional (CH₂)₂ repeat units (m = 2,4,6 and 8). To keep the notation uncluttered, the CH bonds in the Cp rings of the ferrocenyl (Fc) D/A groups have been omitted.

Diabatic States and Properties

The quantities H_{DA} and r_{DA} refer to initial and final states in a charge transfer process, in which the transferring electronic charge is primarily localized, respectively, on the D and A sites (4). We now seek a formal definition of these so-called diabatic (d) states, denoted, respectively, as ψ_A^d and ψ_A^d. It is convenient to achieve this goal by exploiting the relationship between the charge-localized, non-stationary diabatic states of a given system and the corresponding spectroscopic eigenstates, which involve some degree of delocalization of the transferring charge, and are denoted as the adabatic (a) states. Viewing the latter as a basis set, it is necessary to ascertain the smallest set capable of adequately representing the desired diabatic states, ψ_D^d and ψ_A^d (in an overall sequential et process, a corresponding sequence of ψ_D^d, ψ_A^d pairs are required). To proceed further we adopt as a criterion for defining the diabatic states, the maximum degree of charge localization (12,13). Hence for any given space of electronic states (defined, for example, in terms of a set of n adiabatic states), the diabatic states are those which diagonalize the projection of the electronic dipole moment vector operator $(\bar{\mu})$ in the direction of the charge transfer process, and hence maximize the differences among the state dipole moments (projected on the same reference direction). When $n > 2$, one obtains additional (e.g., intermediate) diabatic states, as well as the primary D/A pair. In the traditional Mulliken-Hush (MH) approach (14), it is assumed that a 2-state adiabatic space $\left(\psi_1^a \text{ and } \psi_2^a\right)$ is an adequate basis for representing ψ_D^d and ψ_A^d. The 2-state model is illustrated in Figure 3, which displays initial and final state energy profiles along a reaction coordinate, comparing thermal et occurring between resonant initial and final diabatic states at the transition state (diabatic crossing point) with optical et occurring between initial and final adiabatic states at the ground state equilibrium point, with energy conservation maintained by photon energy $h\nu$. The diabatic and adiabatic energy profiles will in general be distinct due to quantum mechanical mixing, but this feature has been suppressed for simplicity in the schematic depiction in Figure 3.

202

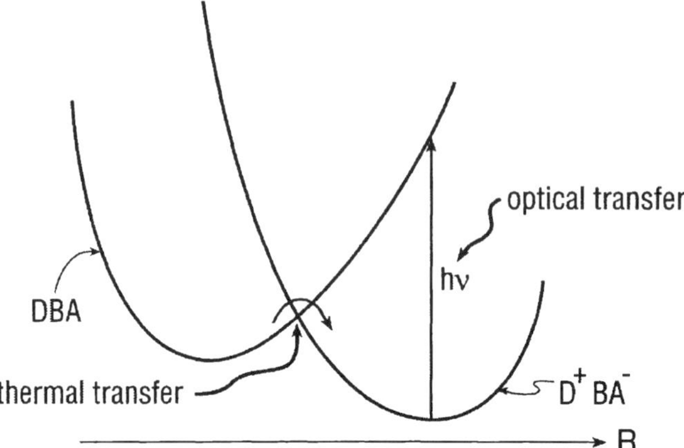

Figure 3. Schematic representation of optical and thermal et, corresponding, respectively, to the vertical transition with excitation energy hv and passage through the transition-state (or crossing) region. In experimental studies, the thermodynamically stable state, which is the final state in the thermal et process, generally serves as the initial state in the corresponding optical process.

Adopting the 2-state approximation and the above criterion for defining diabatic states, together with the assumption that ψ_D^d and ψ_A^d are orthonormal, and the identification of the reference charge transfer direction as that defined by the adiabatic dipole moment shift,

$$\Delta\vec{\mu}_{12}^a \equiv \vec{\mu}_{22}^a - \vec{\mu}_{11}^a \tag{3}$$

where $\vec{\mu}_{ij}^a \equiv \int \psi_i^a\, \vec{\mu}\, \psi_j^a\, d\tau$, leads finally to the following expression for D/A coupling and separation distance according to the Generalized Mulliken Hush (GMH) model (12,13),

$$|H_{DA}| = |\mu_{12}^a\, \Delta E_{12}^a\, /\, \Delta\mu_{DA}^d| \tag{4}$$

where

$$\Delta\mu_{DA}^{d} = \left(\left(\Delta\mu_{12}^{a} \right)^{2} + 4 \left(\mu_{12}^{a} \right)^{2} \right)^{1/2} \tag{5}$$

Here $\Delta E_{12}^{a} \equiv E_{2}^{a} - E_{1}^{a}$ is the adiabatic energy gap (the vertical separation of the 2-state energy eigenvalues). The natural definition of r_{DA} now becomes (12,13),

$$r_{DA} \equiv \Delta\mu_{DA}^{d} / e \tag{6}$$

where e is the magnitude of the electronic charge. In the case of symmetrically delocalized adiabatic states, which are composed of 50-50 mixtures of resonant diabatic states, eq (4) is seen to yield the familiar energy splitting expression for H_{DA} (2,4):

$$| H_{DA} | = | \Delta E_{12}^{a} / 2 | \tag{7}$$

A great advantage of the GMH model (eqs (4) and (5)) is that it can be applied to an arbitrary non-resonant situation. Of course, in some cases it may be necessary to relax the constraints of the 2-state model (13).

Eqs (4) and (5) show that H_{DA} and r_{DA} may be determined entirely on the basis of spectroscopic (i.e., adiabatic state) information, whether from experiment (optical transition energies, oscillator strengths, and dipole shifts from Stark data) or from quantum calculations (self consistent field (SCF) or configuration interaction (CI)). In contrast to the earlier MH model (14), eqs (5) and (6) reveal that the 2-state model implies a value for the effective D/A separation, r_{DA}, without requiring the use of independent assumptions.

In summary, the thermal and optical processes are seen to give complementary accounts of the underlying D/A interaction. Eq (4) shows explicitly how the diabatic *Hamiltonian* matrix element governing the magnitude of the *thermal et* rate constant is related to the adiabatic *dipole* matrix element (the transition dipole), governing the intensity of the *optical et* process linking the same pair of D/A sites.

It has been noted (15,16) that in some cases H_{DA} magnitudes obtained from optical data via eq (4) seem to be underestimated in comparison with independent estimates of H_{DA} from thermochemical data. Such discrepancies may be due to the use of inappropriate estimates of r_{DA} (e.g., using assumed values based on molecular coordinates instead of eq. (5)) or from inadequacies of the 2-state model (16). In any case, ambiguities pertain to all models for estimating H_{DA} magnitudes. In the present study we employ the 2-state GMH model, and in the cases where contact with experiment is possible, reasonable agreement is found. In cases of quasidegeneracy within initial or final state manifolds, where the overall electronic space involves more than two states, it is still found that viable 2-state subspaces may be identified, as discussed further below.

Examples of r_{DA} values are presented in Table I for some transition metal complexes. Not only are the r_{DA} values based on eqs (5) and (6) seen to be appreciably reduced relative to estimates based on molecular geometry (metal atom sites and ligand midpoints), but the distinction between r_{DA} values based on adiabatic $(\Delta\mu_{12}^{a})$ and diabatic $(\Delta\mu_{DA}^{d})$ dipole moment shifts is also clear, so that even in the limit of a fully delocalized system (the Ru(pz)Ru case), where $\Delta\mu_{12}^{a}$ is essentially zero, a finite r_{DA} value is obtained from eqs (5) and (6).

Wavefunction Model

The detailed calculations were carried out for charge shift processes (a variant of the charge separation displayed in eq 1 and Figures 1 and 3) of the type,

$$D^{\pm}BA \rightarrow DBA^{\mp} \tag{8}$$

in which the initial D group is charged.

Eqs (4) and (5) can be implemented in terms of any computational model which yields the necessary energy and dipole moment quantities (these quantities can of course also be obtained from experimental spectroscopic data, as noted above (see Table I)). In the present study we employ orbitals based on state-averaged 2-configuration self-consistent-field (SCF) calculations (13,17). The necessary "spectroscopic" states are then obtained from a 2-state

Table I. Effective Separation of D/A Sites $\left(r_{DA}\left(\mathring{A}\right)\right)$

et System		From Molecular Geometry[a]	From 2-State Analysis of Spectral Data[b]	
			diabatic	adiabatic
$(NH_3)_5\ (Ru^{2+}L)^c$				
MCLT	$L = pz$	3.5	2.2	1.0
	$L = PzH^+$	3.5	2.1	<0.1
	$L = bpy$	5.6	3.4	2.9
	$L - bpyH^+$	5.6	4.3	3.6
$(NH_3)_5\ (Ru^{2+}L\ Ru^{3+})\ (NH_3)_5{}^c$				
IT	$L = pz$	6.8	1.4	<0.1
	$L = bpy$	11.3	5.2	5.1

a)Based on separation of M and ligand (L) midpoint (for MLCT cases) or $r_{MM'}$ (for IT cases), where M and M′ denote Ru atom sites. The notation MLCT and IT corresponds, respectively, to metal-to-ligand charge transfer and intervalence transfer ($\underline{4}$).

b)Based on spectral data analyzed in terms of the 2-state GMH model (eqs. (4) and (5)). In the analysis a value of f = 1.3 was used to relate the local to the applied external electric field in the Stark measurements $\left(\vec{E}_{loc} = f\ \vec{E}_{ext}\right)$.

c)Refs. (31-33); pz $\equiv$ pyrazine; bpy $\equiv$ 4,4-bipyridine.

206

configuration interaction (CI) calculation (17). In the cases dealt with below, the initial and final states are radical ions, either cation (for primarily hole transfer) or anion (for primarily electron transfer), corresponding to charge-shift processes of the generic type depicted in eq 8. The two configurational basis functions in the CI calculations consist of an invariant core together with an active space of 2 orbitals containing 1 (hole transfer for D/A = CH_2) or 3 (all other cases) electrons (as noted above, the orbitals are based on the state-averaged 2-configuration SCF calculation).

The calculations employ the spectroscopic INDO/1 model (INDO/s) developed by Zerner and colleagues (18). This all-valence-electron model has proven to be an extremely powerful tool in evaluating and analyzing H_{DA} quantities for *et* in many types of molecular systems, including both organic and inorganic assemblies (2,4,7,11,17). Comparison with available *ab initio* results has indicated that H_{DA} values from corresponding INDO/s and *ab initio* calculations are often in agreement to within $\pm 25\%$ (17,19).

DBA Structures

The relevant molecular structures are displayed in Figure 2. In the OPE bridges, the ethnylene and para carbon atoms of the phenylenes constitute a straight-line axis, while the phenylene torsion angles ($\{\theta\}$) were allowed to vary. Other details of the structures are given in (17). The OV and OPV bridges were assigned all-trans planar conformations, with values of 1.34 Å, 1.40 Å, and 1.48 Å, respectively, for the double, phenylene, and single C-C bond lengths, and 120° for (idealized) bond angles. In some cases, both ortho hydrogens in the first phenylene of each OPV bridge were replaced with CH_3O-groups. For the OM bridges, C-C bond lengths of 1.53 Å and tetrahedral bond angles were employed, and the terminal carbons were linked to the CH_2 D/A groups with C-C bond lengths of 1.50 Å. D and A groups were taken as trigonal (sp^2-type) CH_2 groups (for all four bridge types) or ferrocenyl (Fc) groups (for OPE and OPV bridge types). For OM bridges, the CH_2 groups were perpendicular to the carbon framework, while for OPV spacers, all atoms were coplanar. When Fc D/A groups were employed, the tethered cyclopentadienyl (Cp) rings and the OPV and OPE bridges were all coplanar. Except as noted, the Fc groups were assigned a cis conformation relative to the plane of the bridge.

Results and Discussion

Decay Coefficients

The main body of calculated results is displayed in Table II as the decay coefficients (β, as in eq (2)) for a variety of DBA systems. Since a primary motivation for the calculations was experimental kinetic data based on a ferrocene/ferrocenium (Fc/Fc^{+}) redox couple tethered to a self-assembled monolayer (SAM)-film-modified gold electrode (6-9), the calculations were carried out mainly for radical cations, $(DBA)^{+}$, where hole transfer may be expected to be the dominant charge transfer mechanism (however, see related comments in (11)). Some comparisons with corresponding radical anions $((DBA)^{-})$ are also provided. The chosen DBA systems employ the same basic oligomeric units as in the systems studied experimentally, but may be viewed as simplified models regarding the D and A moieties. Whereas in the electrochemical kinetic studies, the D and A groups correspond to a gold electrode/Fc pair poised to yield effectively thermoneutral *et* (i.e., with the electrode potential near or equal to the redox potential $\left(E^{o'}\right)$ of the Fc/Fc^{+} couple), the DBA systems used here (see Figure 2) have symmetric D/A pairs (either CH_2 or Fc), thus also corresponding to thermoneutral *et*. Of course, the effective energy gaps separating D/A and bridge levels for the model DBA systems are not the same as those for the actual interfacial systems. An example of a tight binding calculation, which included the gold band structure (as represented by a cluster model) as well as the Fc group and an OM bridge, was reported in (20), with β values similar to those listed in Table II for OM. In addition to comparisons with interfacial thermal kinetic data, the present calculations also include Fc/Fc^{+} linked by OV tethers, allowing comparison with optical *et* data in liquid solution. As in an earlier study, the present calculations employed a Fc molecular structure chosen to correspond to the inner-sphere transition state geometry of the Fc/Fc^{+} redox couple (11).

The omission of solvation in the current *in vacuo* calculations will influence the effective gaps controlling superexchange coupling. The model structures

Table II. Sensitivity of H_{DA} decay to D/A and B type in $(DBA)^{\pm}$ systems[a]

_D/A_	_B (conformation)_	_Calculated_		_exp_
		radical cations	_radical anions_	
CH_2	OV^b (planar)	0.31	0.32	$\geq 0.2^c$
CH_2	OPE^d (planar)	0.39	0.43	$0.4^{f,g}, 0.6^{f,h}$
	OPE^d (randomized)	0.51	0.54	
Fc^e	OPE^d (planar)	0.36	$--$	
CH_2	OPV^i (planar)	0.43	0.46	$--^{f,k}$
	OPV^i (planar) – CH_3O-subst.)^j	0.46	$--$	
Fc^e	OPV^i (planar)	0.42	$--$	
CH_2	OM^l (staggered)	0.83	1.00	$0.9\pm0.1^{f,m}$

a)Detailed structures are given in Figure 2. The results for OV, OM, and some of the OPE and OPV cases have been reported in (7), (9), and (17).

b)OV ≡ oligovinylene

c)Refs. (27,34,35).

d)OPE ≡ oligo-p-phenylenevinylene

e)Hole ground states are localized in the $3d_{x^2-y^2}$ Fe orbitals, where z is the Fc (ferrocenyl) axis and x is long axis of the bridge.

f)Based on interfacial kinetic measurements at film-modified Au electrodes.

g)Ref. (8)

h)Ref. (7)

i)OPV ≡ oligo-p-phenylenevinylene

j)The two ortho hydrogens nearest to the donor attachment site of the first phenylene group were replaced with methoxy groups.

k)Ref. (9)

l)OM ≡ oligomethylene

m)Ref. (6)

also differ from the experimental systems in some other details (e.g., the inclusion of an extra ethynylene or vinylene unit in the OPE and OPV bridges, respectively (see Figure 2), and the degree of substitution of the phenylene groups). Finally, we note that the present study by construction deals only with electronic coupling mediated by the oligomer tethered to the redox group, thus omitting any contributions involving inter-chain contacts (the possibility of such effects has been discussed in (21)).

In spite of the limitations noted above, the model systems yield decay patterns (Table II) in good overall accord with available experimental data. Furthermore, the oversimplified CH_2 model for D and A groups is seen to yield results quite close to those based on the Fc D and A groups. Finally, we find that corresponding radical cation and anion pairs yield very similar β values. The β values in Table II represent mean values over the range of oligomeric systems defined in Figure 2, i.e., spanning r_{DA} magnitudes up to ~30Å There is of course no reason for the calculated results to conform exactly to a simple exponential falloff (eq 2)), but in fact the calculated behavior is nearly exponential (essentially exact for the OM case). Defining a "local" β, based on adjacent pairs of oligomers (m, m+1), we note that those quantities *increase* with *m* for CH_2 D/A groups (with an overall range of 0.04 Å^{-1}), and *decrease* with m for Fc D/A groups (with overall range of 0.08 Å^{-1} (OPE) and 0.04 Å^{-1} (OPV)).

Table II reveals a number of important trends. Within the set of systems involving planar unsaturated bridges (all-trans or fully-extended) the most gradual decay is exhibited by vinylene units (OV), with somewhat steeper falloff when phenylene units are present, moreso in combination with vinylenes (OPV) than with ethnylenes (OPE). While alkoxy substitution of a phenyl group is known to raise the occupied orbital energy levels (e.g., see (22)), as also found in the present calculations, and thus might be expected to enhance hole tunnelling (and reduce β) by reducing the relevant D/A - B energy gap (2), the net result from the detailed calculations is seen (Table II) to be only a small change in β, but one which is positive. Detailed analysis of superexchange tunnelling has shown that the overall process may involve a large number of superposed "pathways", so that the response to the change of a given gap may be quite complex (23). While departure from planarity may be significant for coupling mediated by any of the unsaturated bridges, the issue is especially important for the OPE bridges (7,17) in view of the small intrinsic barrier (~k_BT

at room temperature) to torsional motion of the phenylenes (denoted by θ in Figure 2). In view of the lack of detailed information regarding torsional fluctuations, we have evaluated the coupling for thermally averaged torsional motion (either fully random, or subject to the barrier noted above, imposed for each phenylene unit). The result (17) is an appreciably increased β value (0.51 Å^{-1} for the radical cation), which is closer to the experimental estimate (0.6 Å^{-1}, based on m=2 and m=3 (7)) for unsubstituted OPE's than the calculated result based on planar OPE's (0.39 Å^{-1}). On the other hand, the experimental β value obtained for substituted OPE's (0.4 Å^{-1} (8)) is quite close to the calculated result for planar OPE's. Since the limited set of calculations reported here for methoxy substituent effects indicate only a small, positive effect on beta, it may well be that the substituents in the experimental systems are indirectly affecting the coupling by influencing the energetics of inter chain contacts, which are important factors in determing the torsional configuration within a given OPE chain (cf. (24)).

It should be emphasized that β as defined here (eq (2)) applies strictly only to the distance dependence of the square of the electronic coupling factor, H_{DA}, and its utility is basically limited to the regime of sufficiently large energy gaps (D/A *vs* B), where superexchange tunnelling pertains. Its relationship to the distance dependence of the full rate constant requires knowledge of other sources of distance dependence (mainly, the activation energy) and of the overall reaction mechanism. The calculations reported here yield similar β values for comparable (i.e., planar) OPE and OPV bridges. In contrast, the observed rate constants for OPV systems(9) tend to be flatter with respect to *m* than for similar OPE systems(7,8). Even within the class of OPV systems, however, the details of the variation with *m* of the rate constant depend on the specific details of the DBA system (cf. (9) and (25)). Also, we have noted above the apparent dependence of experimental β values for OPE systems on phenylene substituents (the experimental estimates of β listed in (7) and (8) are based on the rate constants).

Comparison of Calculated and Experimental H_{DA} Magnitudes

In Table III results are presented which allow a comparison of calculated and experimental H_{DA} magnitudes for (Fc OPE (0) Fc)$^+$ and (Fc OPV (0) Fc)$^+$

systems (26,27). Here, aside from a single CH_3O- substituent on each Fc in the OPE (0) system studied experimentally (26), the chemical structures used in the calculations and in the experiments are the same. There is, however, an ambiguity regarding the conformation of the Fc groups with respect to the bridges in the systems studied experimentally (trans and cis limiting conformations were considered). All of the different H_{DA} magnitudes are similar (a few hundred cm^{-1}). The calculated results show two different trends: OPE (0) > OPV (0) for the trans conformation (by ~ 20%); and trans > cis (for OPE (0), but only by 5%. The experimental data was analyzed in terms of the Mulliken-Hush (MH) model (eq (4)), using the spectroscopic values of ΔE_{12}^a and μ_{12}^a (the latter is inferred from the extinction coefficient maximum and the line width), with r_{DA} estimated as r_{FeFe} based on structural models for the cis and trans isomers. Our calculations for OPE (0) indicate that μ_{12}^a and $\Delta \mu_{DA}^d$ (and hence r_{DA}) have very similar trans/cis ratios, with the overall trans/cis H_{DA} ratio being dominated by the trans/cis ratio of ΔE_{12}^a values (1.05). Hence the calculated H_{DA} (trans) slightly exceeds H_{DA} (cis), even though the r_{DA} values by themselves (via eq (5)) yield a ratio less than unity (0.9).

Alternative hole states

The (Fc OPE (m) Fe)$^+$ and (Fc OPV (m) Fc)$^+$ results in Tables II and III are based on ground-state radical cation states, in which the holes are mainly localized in 3d Fe atomic orbitals of the x^2-y^2 type where z is the Fc axis, perpendicular to the Cp rings, and the bonds linking the Fc's to the bridge are parallel to the x axis.

In the isolated ferrocene, the $3d_{x^2-y^2}$ and $3d_{xy}$ orbitals (and hole states) are degenerate, but due to different orientation of nodal planes in the tethered DBA systems, $3d_{x^2-y^2}$, which overlaps with a Cp molecular orbital (MO) having a finite contribution from the carbon atom linked to the bridge, lies above the $3d_{xy}$, orbital, which has a node at the linked Cp carbon atom. Thus the hole ground states ($3d_{x^2-y^2}$) are the ones which yield good overall D/A coupling, as displayed in Table IV. Nevertheless, the $3d_{xy}$ hole states, which yield very small

Table III. Comparison of Calculated and Experimental H_{DA} Values for (Fc OPE (0))$^+$ and (Fc(OPV (0) Fc)$^+$

System	r_{FeFe} $(Å)^a$	H_{DA} (cm^{-1})	
		calculated[b]	experiment[c]
(Fc OPE(0)Fc)$^+$	6.4 (cis)	649(cis)	525 (cis r_{FeFe})[d]
	7.3 (trans)	682(trans)	427 (trans r_{FeFe})[d]
(Fc OPV(0)Fc)$^+$	6.2 (cis)	589 (cis)[e]	559 (cis r_{FeFe})[f]
			492 (trans r_{FeFe})[f]

a)Values for the structures used in the present calculations.

b)Based on the GMH model (eqs (4) and (5)).

c)Based on the MH model, taking r_{DA} as the r_{FeFe} estimates for cis and trans isomers.

d)Ref. (26); results based on acetonitrile solvent; cis and trans r_{FeFe} values taken, respectively, as 5.85 and 7.20 Å.

e)Obtained from the calculated Fc OPV (1)Fc)$^+$ value and $\beta = 0.44$ Å^{-1} (based on the m= 1 and m=2 systems). The mean value of $\beta = 0.42$ Å^{-1} (Table II) yields 552 cm^{-1}.

f)Ref (27); results based on dichloromethane solvent; cis and trans r_{FeFe} values taken as 6.19 Å and 7.03 Å.

Table IV. Quasi degenerate hole states in (Fc OPE(1) Fc)$^+$ and (Fc OVP(1)Fc)$^+$

	Fe 3d-hole type[a]	H_{DA} (cm^{-1})[b]	Relative Energy (eV)[c]
(Fc OPE(1)Fc)$^+$	$3d_{x^2-y^2}$	177	0.000
	$3d_{xy}$	2	0.058
(Fc OPV (1)Fc)$^+$	$3d_{x^2-y^2}$	140	0.000
	$3d_{xy}$	0.4	0.051

a)Dominant character of calculated hole state where z is the Fc axis and the x axis is aligned with the single bonds linking the Fc Cp rings to the bridge. In the $3d_{xy}$-hole states, the hole resides predominantly in an Fe orbital which mixes with Cp molecular orbitals bearing a node on the carbon atom linked to the bridge, thus accounting for the very small H_{DA} magnitudes.

b)The β values in Table II are based on ground-state ($3d_{x^2-y^2}$-type) hole states.

c)Based on the SCF/CI calculations described in the text.

coupling ($|H_{DA}| \leq 1$ cm^{-1}), are calculated to lie only .05-.06 eV higher and thus are thermally accessible at room temperature. The relative occupations of the two hole states could of course be altered by suitable chemical substituents on the Cp rings.

Energy gaps for Hole Transfer into the Bridge

A crucial quantity controlling the nature of D/A coupling and charge transport is the energy gap for transferring ("injecting") charge from D to the bridge (1-5). Table V displays some gaps obtained from CI calculations for (DBA$^+$) systems containing the OPE (1) or OPV (1) bridge. The model D/A CH_2 groups are seen to yield a gap appreciably greater that the gap based on Fc groups. The OPE and OPV bridges yield very similar gaps for a given D/A pair.

The calculated *in vacuo* values, which are comparable to, but somewhat larger than, corresponding estimates based on redox potentials in polar organic solvents (e.g., using data given in (7,9,25)), are large enough to place the D/A coupling well within the superexchange tunnelling regime, at least for m=1, and the observed β values (Table II) indicate that the higher homologs are also within the superexchange regime, even though the gaps are expected to be somewhat reduced with increasing *m* (25). As already noted, however, the electronic tunnelling step does not appear to be rate determining in several cases of *et* involving OPV (m) bridges.

Table V. Calculated Energy Gaps for injecting holes into OPE(1) and OPV(1) bridges $\left(D^{+}BA \rightarrow DB^{+}A\right)$

D/A	*Bridge*	$\Delta E^{a}(eV)$
Fc	OPE(1)	1.6
CH$_2$	OPE(1)	2.4
Fc	OPV(1)	1.5
CH$_2$	OPV(1)	2.3

a)Estimates based on CI calculations including single excitations from the occupied π-type MO's predominately localized on the bridge into the singly-occupied MO of the doublet hole ground state.

A cobalt-based analog

Michl and coworkers have discussed the potential role of $(C_4R_4)Co(C_5H_5)$ complexes (Figure 4) in conductive 2-dimensional assemblies, in which the cyclobutadiene rings may be linked via suitable substituents R (28). Denoting the complex as CbCoCp, where R=H for the present discussion, we note that it may be viewed as basically isoelectronic (as far as the relevant Co and ligand MO's are concerned) with ferrocene (CpFeCp), having nominal group oxidation states Cb^{2-} Co^{3+} Cp^{-} in contrast to Cp^{-} Fe^{2+} Cp^{-}. In Table VI we compare the low-lying hole states of CbCoCp and CpFeCp, based on open-shell restricted

Hartree Fock calculations. The primary feature of interest is the relative order of the non-degenerate $3d_{x^2-y^2}$ and $3d_{xy}$ hole states in $(CbCoCp)^{+}$. While the overall symmetry of the CbCoCp structure can be at most C_s, the local C_{4v} symmetry of the CbCo moiety plays an important role, with the anitbonding Cb pi MO (b_1) mixing with the Co $3d_{x^2-y^2}$ orbital and pushing it below the $3d_{xy}$ orbital (b_2), whose nodal planes contain the Cb carbon atoms and in the local C_{4v} symmetry do not mix with any of the Cb pi MO's. This simple orbital argument can explain qualitatively why the ground hole state is $3d_{xy}$ (29). This result has implications for electronic coupling when CbCoCp units are linked, since as discussed above, the $3d_{xy}$ orbitals yield much less favorable coupling than the $3d_{x^2-y^2}$ orbitals. On the other hand, covalent coupling to a bridge will tend to raise the $3d_{x^2-y^2}$ energy level relative to that for $3d_{xy}$ (as happens when

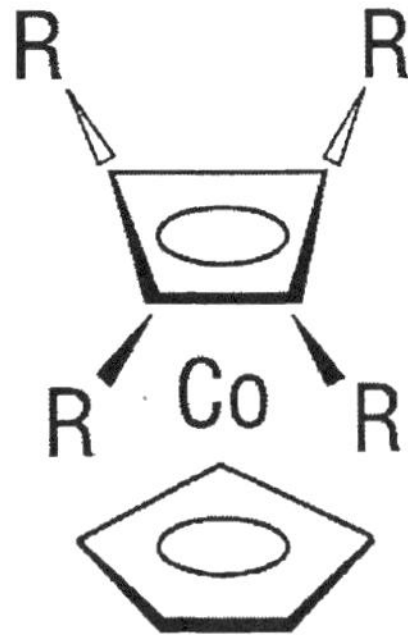

Figure 4. Cyclobutadiene (R = H)/cyclopentadiene Co sandwich complex (isoelectronic with ferrocene as far as relevant valence molecular orbitals are concerned).

Table VI. Comparisons of Low-Lying Hole States for Ferrocene and $(C_4H_4)Co(C_5H_5)$[a]

	Hole State[b]	*Relative Energy (eV)*[c]
Ferrocene	$3d_{x^2-y^2}$, $3d_{xy}$	0.00
	$3d_{z^2}$	0.48
$(C_4H_4)Co(C_5H_5)$[a]	$3d_{xy}$	0.00
	$3d_{z^2}$	0.27
	$3d_{x^2-y^2}$,	0.33

a)Structure given in Figure 4 (R=H).

b)The 3d-type hole states are based on a coordinate system where the z axis is perpendicular to the ring planes and where the x and y axes are parallel to the diagonals of the C_4 ring. Thus the C_4 carbons lie in the nodal planes of the $3d_{xy}$ orbital (cf related comments in footnote *a* of Table IV).

c)The ground and low-lying excited states were obtained from separate direct SCF calculations. The different states are symmetry-distinct for the assumed D_{5h} ferrocene geometry. For (C_4H_4) Co (C_5H_5), distinct SCF solutions for $3d_{x^2-y^2}$, and $3d_{z2}$ hole states converged cleanly even though strictly speaking, the relevant 3d orbitals both transform as A′ in the assumed C_s point group adopted for the structure (in the local C_{4v} symmetry of the (C_4H_4) Co moiety, the two orbitals transform as b_1 and b_2).

tethers break the $3d_{x^2-y^2}/3d_{xy}$ degeneracy in the case of Fc-based DBA systems). Substituents (R) on the Cb ring will also have a bearing on the relative $d_{x^2-y^2}$ and d_{xy} energy levels. The D/A coupling in tethered CbCoCp systems is the subject of ongoing investigation (30)

Conclusions

Electronic coupling matrix elements (H_{DA}) elements for a number of radical cation and anion systems of type $DBA^{\pm}$, where D,B, and A denote donor, bridge and acceptor moieties, have been evaluated and analyzed, using configuration interaction electronic structure calculations (INDO/s) in conjunction with the Generalized Mulliken Hush model, which also yields effective D/A separation distances (r_{DA}). The systems studied include D/A = CH_2 or ferrocenyl, and B = oligovinylene (OV), oligo-p-phenylene-ethynylene (OPE), oligo-p-phenylene-vinylene (OPV), and oligomethylene (OM). Coupling based on alternative hole-states has been considered and electronic structural features relevant to D/A coupling have been compared for ferrocene and a cobalt-based analog.

Nearly exponential decay of H_{DA} with respect to r_{DA} is observed in all cases, with the mean decay coefficient (β) for all-planar conformations increasing in the order OV < OPE $\leq$ OPV < OM. The β value for OPE increases appreciably when account is taken of torsional fluctuations, while methoxy substitution of the OPV phenylene groups causes only a small increase. The calculated β values for OV, OPE, and OM bridges are in reasonable agreement with experimental estimates. The similarity of β values for planar OPE and OPV bridges is quite distinct from the pronounced difference in the distance dependence of the corresponding experimentally observed thermal rate constants, thus suggesting that different rate determining steps may be controlling the kinetics of the respective electron transfer processes. For a given bridge, calculations based on CH_2 or Fc D/A groups yield quite similar β values. Consistent with superexchange tunnelling, substantial energy gaps for hole injection from D to B are calculated ($\geq$ 1.5 eV). Low-lying thermally-accessible excited 3d-hole states at the Fc donor are found to yield small H_{DA} magnitudes as a result of orbital nodal properties. An inversion of the two types of hole state is calculated to occur in a Co analog of ferrocene, in which a cyclobutadiene (Cb) ring replaces a cyclopentadiene ring, but the quantitative details are expected to be sensitive to chemical substitution on the Cb ring.

Acknowledgement. This research was carried out at Brookhaven National Laboratory under contract DE-AC02-98CH10886 with the U.S. Department of Energy and supported by its Division of Chemical Sciences, Office of Basic Energy Science. We benefited from stimulating discussions with Drs. J. P. Kirby and J. R. Miller.

References

1. Jortner, J.; Bixon, M. *Adv. Chem. Phys.* **1999**, *106*, 1-734.
2. Newton, M. D. *Chem. Rev.* **1991**, *91*, 767-792.
3. Jortner, J.; Ratner, M. A. In *Molecular Electronics*; Jortner, J.; Ratner, M., Eds.; *Blackwell Science*: Oxford, **1997**, pp 5-55.
4. Newton, M. D. In *Electron Transfer in Chemistry*, Balzani, V., Ed.; Wiley-VCH: Weinheim, **2001**, Vol. 1, pp. 3-63.
5. Segal, D.; Nitzan, A.; Davis, W. B.; Wasielewski, M. R.; Ratner, M. A. *J. Phys. Chem. B* **2000**, *104*, 3817-3829.
6. Smalley, J. F.; Feldberg, S. W.; Chidsey, C. E. D.; Linford, M. R.; Newton, M. D.; Liu, Y.-P. *J. Phys. Chem.* **1995**, *99*, 13141-49.
7. Sachs, S. B.; Dudek; S. P.; Hsung, R. P.; Sita, L. R.; Smalley, J. F.; Newton, M. D.; Feldberg, S. W; Chidsey, C. E. D. *J. Am. Chem. Soc.* **1997**, *119*, 10563-64.
8. Creager, S.; Yu, C. J.; Bamdad, C.; O'Connor, S.; MacLean, T.; Lam, E.; Chong, Y.; Olsen, G. T.; Luo, J.; Gozin, M.; Kayyem, J. F. *J. Am. Chem. Soc.* **1999**, *121*, 1059-1064.
9. Sikes, H. D.; Smalley, J. F.; Dudek, S. P.; Cook, A. R.; Newton, M. D.; Chidsey, C. E. D.; Feldberg, S. W. *Science* **2001**, *291*, 1519-23.
10. Jortner, J.; Bixon, M. *J. Chem. Phys.* **1988**, *88*, 167-70.
11. Newton, M. D.; Ohta, K.; Zhong, E. *J. Phys. Chem.* **1991**, *95*, 2317-26.
12. Cave, R. J.; Newton, M. D. *Chem. Phys. Lett.* **1996**, *249*, 15-19.
13. Cave, R. J.; Newton, M. D. *J. Chem. Phys.* **1997**, *106*, 9213-9213.
14. Hush, N. S. *Electrochim Acta* **1968**, *13*, 1005-23.
15. Salaymeh F.; Berhane, S.; Yusof, R.; de la Rosa, R.; Fung, E. Y.; Matamoros, R.; Lau, K. W.; Zheng, Q.; Kober, E. M.; Curtis, J. C. *Inorg. Chem.* **1993**, *32*, 3895-3908.
16. Dong, Y.; Hupp, J. T. *Inorg. Chem.* **1992**, *31*, 3170-72.
17. Newton, M. D. *Int. J. Quant. Chem.* **2000**, *77*, 255-263.

18. Zerner, M. C.; Loew, G. H.; Kirchner, R. F.; Mueller-Westerhoff, U. T. *J. Am. Chem. Soc.* **1980**, *102*, 589-99.

19. Newton, M. D. In *Cluster Models for Surface and Bulk Phenomena*, Pacchioni, G.; Bagus, P. S.; Parmigiani, F., Eds.: Plenum Press, New York, **1992**, pp. 551-63.

20. Hsu, C.-P.; *J. Electroanal. Chem.* **1997**, *438*, 27-35.

21. Slowinski, K.; Chamberlain, R. V., Miller, C. J., Majda, M. *J. Am. Chem. Soc.* **1997**, *119*, 11910-19.

22. Davis, W. B.; Wasielewski, M. A.; Ratner, M. A. *Int. J. Quant. Chem.* **1999**, *72*, 463-71.

23. Newton, M. D.; Cave, R. J. In *Molecular Electronics*, Jortner, J.; Ratner, M. A., Eds.; Blackwell Science: Oxford, **1997**, pp. 73-118.

24. Seminario, J. M.; Zacarias, A. G.; Tour, J. M. *J. Am. Chem. Soc.* **1998**, *120*, 3970.

25. Davis, W. B.; Svec, W. A.; Ratner, M. A.; Wasielewski, M. R. *Nature* **1998**, *396*, 60-63.

26. Plenio, H.; Hermann, J.; Sehring, A. *Chem. Eur. J.* **2000**, *6*, 1820-29.

27. Ribou, A.-C.; Launay, J.-P.; Sachtleben, M. L.; Li, H.; Spangler, C. W. *Inorg. Chem.* **1996**, *35*, 3735-40.

28. Harrison, R. M.; Brotin, T.; Noll, B. C.; Michl, J. *Organometallics* **1997**, *16*, 3401-12.

29. Stoll, M. E.; Lovelace, S. R.; Geiger, W. E.; Schimanke, H.; Hyla-Kryspin, I.; Gleiter, R. *J. Am. Chem. Soc.* **1999**, *121*, 9343-51.

30. Newton, M. D. (unpublished work).

31. Reimers J. R., Hush, N. S. *J. Phys. Chem.* **1991**, *95*, 9773-81.

32. Oh, D. H.; Sano, M.; Boxer, S. G. *J. Am. Chem. Soc.* **1991**, *113*, 6880-90.

33. Brunschwig, B. S.; Creutz, C.; Sutin, N. *Coordination Chem. Rev.* **1998**, *177*, 61-79.

34. Woitellier, S.; Launay, J. P.; Spangler, C. W. *Inorg. Chem.* **1989**, *28*, 758-62.

35 Reimers, J. R.; Hush, N. S. *Inorg. Chem.* **1990**, *29*, 3686-97 and 4510-13.

Chapter 16

Electronic Transport in Molecular Devices from First Principles

M. Di Ventra[1], N. D. Lang[2], and S. T. Pantelides[3]

[1]Physics Department, Virginia Polytecnic Institute and State University,
209A Robeson Hall, Blacksburg, VA 24061–0435
[2]IBM Research Division, Thomas J. Watson Research Center, Yorktown Heights, NY 10598
[3]Department of Physics, Vanderbilt University, Nashville, TN 37235

We present an overview of recent work by the authors in first-principles calculations of electronic transport in molecules for which experimental results are available. We find that the shape of the current-voltage characteristics is mostly determined by the electronic structure of the molecules in the presence of the external voltage whereas the absolute magnitude of the current is determined by the chemistry of individual atoms at the contacts. A three-terminal device has been modeled, showing gain. Current-induced forces have been found to induce collective oscillations of the benzene-1,4-dithiolate molecule at resonant-tunneling condition. Finally, we account for recent data that show large negative differential resistance and a peak that shifts substantially as a function of temperature.

Silicon-based microelectronics is reaching the level of miniaturization where quantum phenomena such as tunneling cannot be avoided and the control of doping in ultrasmall regions becomes problematic. Though it is likely that silicon-based technology will simply move to a different paradigm and continue taking advantage of the existing vast infrastructure and manufacturing capabilities, novel and alternative approaches may give new insights and ultimately may usher a new era in nanoelectronics. Molecules as individual active devices are obvious candidates for the ultimate ultrasmall components in nanoelectronics. Though the idea has been around for more than two decades, (*1*) only recently measurements of current-voltage characteristics of individual molecules have been feasible. (*2-4*)

Semiempirical methods for the calculation of current in small structures placed between two metal electrodes have been developed over the years. (*5-9*) For instance, semiempirical tight-binding models have been remarkably successful to provide a useful account of the linear transport properties of molecules and nanotubes. (*5-9*) On the other hand, charge redistribution in the molecules and at the contacts is an important issue to understand, for example, the role of current-induced forces or non-linear effects. In this case a set of first-principles methods is necessary.

In the 1980's, one of us (NDL) developed a practical method to calculate transport in the context of imaging atoms with scanning tunneling microscopy. The method has all the ingredients needed to compute current-voltage (I-V) characteristics of single molecules. (*10*) Recently, two of us (MDV and STP) developed a suitable Hellmann-Feynman theorem for the calculation of current-induced forces on atoms that allows studying the effect of current on atomic relaxations and ultimately the breakdown of molecular devices. (*11*) Combining these tools, we have carried out extensive studies of transport in molecules whose core is a single benzene ring. Such molecules have been synthesized and measured by Reed and coworkers. (*2-4*) In this paper we summarize the most important results of the recent work. The method of calculation and more details of the results can be found in the original papers. (*10-11*)

Transport in the benzene-1,4-dithiolate device

Fig. 1 shows a schematic of a benzene ring making contact to macroscopic gold electrodes via sulfur atoms at each end. The experimental I-V characteristic of a similar device is shown in the top panel of Fig. 2 as measured by Reed *et al.* (*2*). In the middle panel of the same figure we report the theoretical results. We will discuss the third panel shortly. It is clear from figure 2 that the shape of the I-V curve is reproduced quite well, but the absolute magnitude of the current is off by more than two orders of magnitude. We first discuss the origin of the shape of the I-V curve, and then address the issue related to the absolute value of the current. In Fig. 3 we show the density of states of the molecule for three different voltages: 0.01 V, 2.4 V and 4.4 V. We also indicate the energy window between

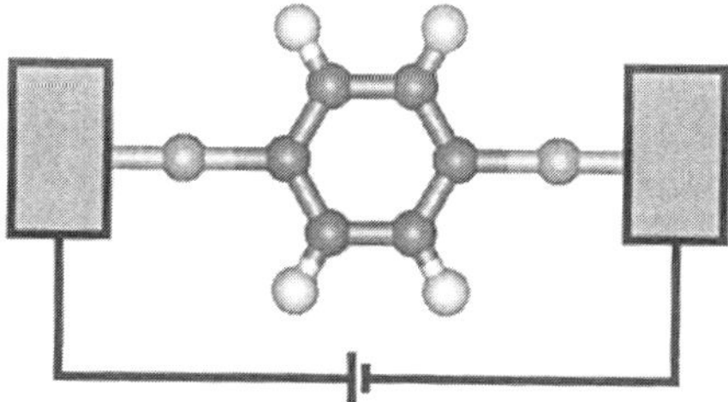

Figure 1: Schematic of a benzene molecule connected to gold electrodes via sulfur atoms.

the left and right quasi-Fermi levels. States within this energy window contribute directly to electronic transport. We see that there is virtually no density of states in the small energy window at small voltages, in agreement with the slow initial rise of the I-V curve. At 2.4 V, the π^* states of the molecule enter the energy window for transport and give rise to the first peak in the spectrum. The peak at 2.4 V is somewhat more pronounced in the theoretical curve. The observed smoothing is likely to be caused by interactions between the electrons and vibrational modes. Finally, at 4.4 V, the π states of the molecule enter the energy window for transport, giving rise to the second peak in the spectrum. At this value of the external potential the π^* continue to participate to transport but at a relatively lower density.

The particular device geometry used in the experiment by Reed *et al.* (2) suggests that the contacts are atomically terminated. In order to explore the mechanism that controls the absolute magnitude of the current we therefore performed calculations by inserting an extra gold atom between the sulfur atom and the macroscopic electrode at each end of the molecule. Indeed, we found a dramatic decrease in the current, bringing its value much closer to the experimental value (see bottom panel of Fig. 2). The decrease is attributed to the fact that gold atoms have only one s orbital available for transport at the Fermi level. However, the gold s orbitals do not couple with the p orbitals of the sulfur atom parallel to the metal surface, while those perpendicular to the metal surface form σ bonds with the gold s orbitals. These σ bonds are, however, located at a much lower energy than the Fermi level. The single-gold-atom-sulfur bond thus breaks the channel for electronic transport. To futher test this idea we performed calculations by replacing the gold atoms with aluminum atoms. The latter have p electrons that should couple well with the p electrons of the molecule. Indeed, the current increases by about an order of magnitude with aluminum atoms replacing the gold atoms. An additional test was carried out with each sulfur atom in front of the center of a triangular pad of three gold atoms on each electrode surface. The current was again at its full value because the three s orbitals on the three gold atoms can form enough linear combinations to produce

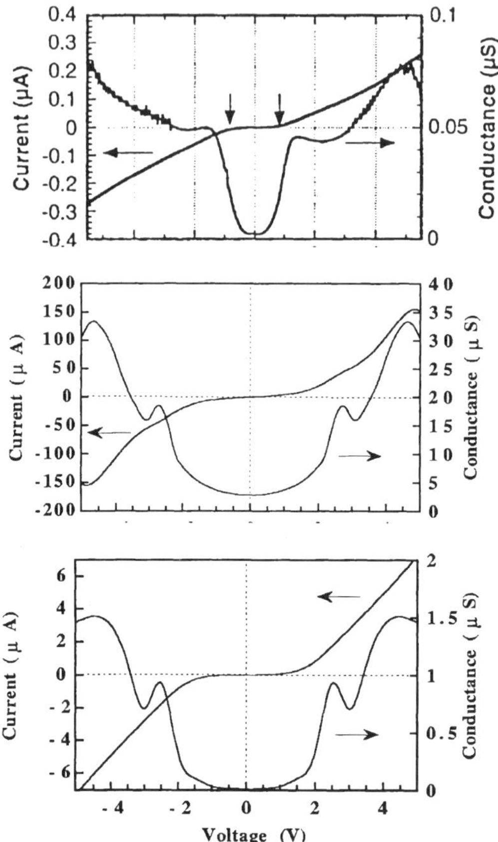

Figure 2: Top panel: experimental I-V characteristics of the benzene-1,4-dithiolate molecule (Ref. (2)). Middle panel: theoretical I-V characteristics of the molecule of Fig. 1. Bottom panel: theoretical I-V characteristics of the molecule of Fig. 1 with one extra gold atom between S and gold electrode at each end (Ref. (12)).

sufficient coupling. (*12*) It is clear from the above that molecules determine the shape of the I-V characteristic, but the nature of individual atoms at the molecule-electrode contact determines the absolute magnitude of the current.

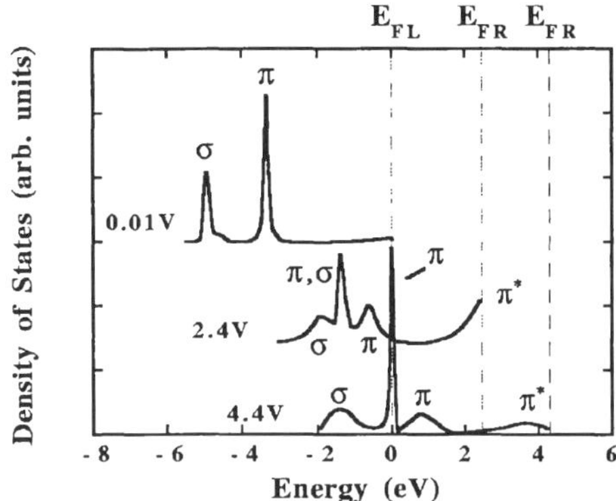

Figure 3: Difference between the density of states of the two semi-infinite electrodes with and without the benzene-1,4-dithiolate molecule in between, for three different voltages. The left Fermi level E_{FL} has been chosen as the zero of energy. The labels E_{FR} correspond to the energy position of the right Fermi levels. The three curves correspond to the bias voltages indicated. (Reproduced with permission from reference 12. Copyright 2000 American Physical Society)

Three-terminal devices

We investigate now the effects of a polarization field in the direction perpendicular to the current flow. (*13*) To this aim, a third terminal is introduced in the form of a capacitor field generated by two circular charged disks at a certain distance from each other placed perpendicular to the transport direction. The disks are kept at a certain potential difference with one of the two disks at the same potential energy as the source Fermi level (see Fig. 4). The axis of the cylindrical capacitor is on the plane of the benzene ring. Since in practical realizations of this device the gate could be of different form and size we discuss the results in terms of applied gate field along the axis of the capacitor. The calculated I-V characteristic as a function of the gate bias is shown in Fig. 5. The source-drain voltage difference has been fixed at 10 mV. We observe an increase of the current with the gate field, after a region of nearly constant current. The current reaches a maximum value at 1.1 V/Å, then decreases till about 1.5 V/Å, to increase further afterwards linearly. The different features of the I-V curve can

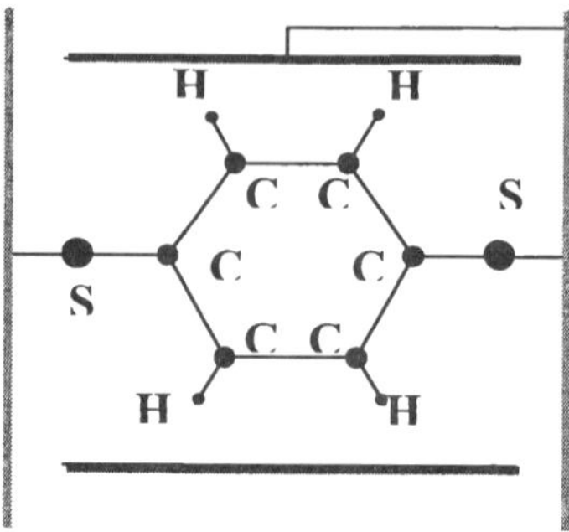

Figure 4: Scheme of the three-terminal geometry used in the present study. The molecule is sandwiched between source and drain electrodes along the direction of electronic transport. The gate electrodes are placed perpendicular to the molecule plane. (Reproduced with permission from reference 13. Copyright 2000 American Institute of Physics)

be understood by looking at the density of states for different gate voltages. The initial slow rise of the conductance represents basically ohmic behavior. It is also observed experimentally for the two-terminal geometry. (*2*) The top curve of Fig. 6 shows the small density of states through which current can flow. The

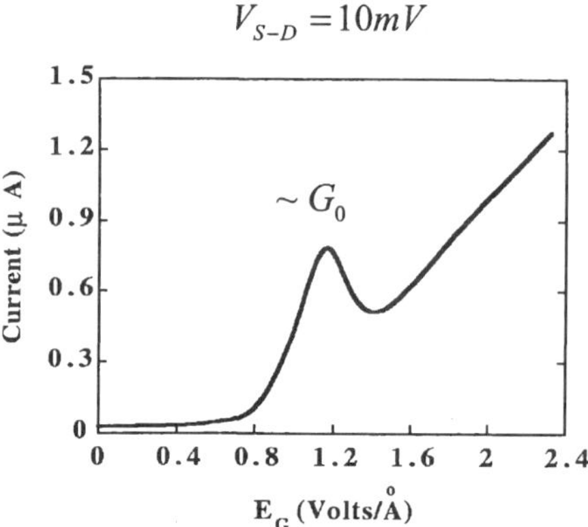

Figure 5: Conductance of the molecule of Fig. 4 as a function of the external gate field.

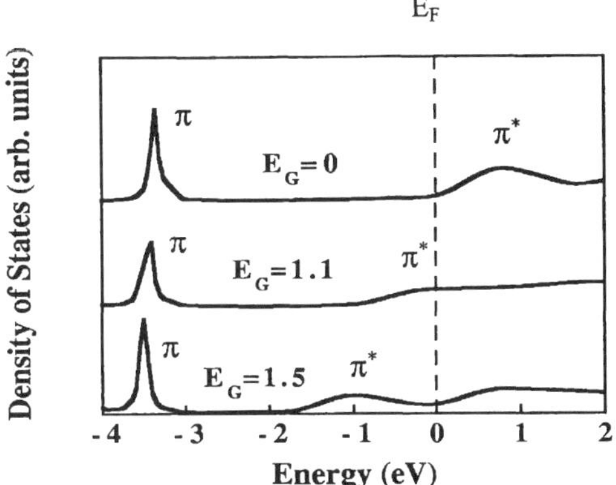

Figure 6: Difference between the density of states of the two semi-infinite electrodes with and without the benzene-1,4-dithiolate molecule in between, for three different gate voltages. The left Fermi level has been chosen as the zero of energy. (Reproduced with permission from reference 13. Copyright 2000 American Institute of Physics)

σ and π bonding states are several eV below the Fermi levels while the π^* antibonding states are nearly 1 eV above the Fermi levels.

A peak and subsequent valley in the current-voltage characteristic are observed at about 1.1 V/Å. From Fig. 6 (middle curve) it is evident that the peak and the valley are due to resonant tunneling through π^* antibonding states. The antibonding states shift in energy with increasing gate bias, and eventually enter into resonance with the states between the right and left Fermi levels, separated by 10 meV (see Fig. 6). The π states are less affected by the gate bias than the π^* states due to a lower density of continuum states in their proximity. Therefore the gap between the π states and π^* states decreases about 1 eV at resonance. Increasing the bias further (bottom curve of Fig. 6), the resonant-tunneling condition is lost and a valley in the I-V characteristic is observed. Finally, as the gate bias is increased further the current starts to increase with the gate bias. The peak-to-valley ratio of the present system ($\sim$ 1.4) is probably too small to be observed at room temperature. This has already been argued in the two-terminal geometry case: the peak-to-valley ratios observed experimentally in the present system for the two-terminal geometry are considerably smaller than theoretically predicted (see discussion above). The value of the gate field at which resonant tunneling occurs (1V/Å) seems slightly high for a molecule of nominal length of 8 Å. Two observations are however in order: i) we have previously shown (*12*) that the theoretical peak of transmission due to antibonding states occurs at higher external bias than the experimental one. ii) In a practical realization of the device the capacitance field would certainly leak into the source and drain electrodes, providing a packet of electrons with higher kinetic energy to tunnel, thus reducing the gate bias value at which the resonance occurs. Both above remarks give us confidence that resonant tunneling and amplification can occur at lower gate field. Current amplification has been actually observed recently by Shon *et al.* in similar devices. (*14*)

Current-induced forces

The effect of forces on the two-terminal molecular device have been investigated in the case of the molecule connected to two bare electrode metals. The current flow has essentially two effects on the atomic structure: i) the molecule twists around the axis perpendicular to its plane, ii) the molecule expands at the bias for which the first resonant tunneling condition occurs, then contracts at about 2.8V corresponding to the valley of Fig. 2, middle panel. The first effect increases with increasing bias even though the global current is not substantially altered. The second effect is due to the deplation of charge between the central C atoms at the first resonance peak due to resonant tunneling with antibonding states. The charge is again recovered in the central C bonds when the resonant tunneling condition is lost and the molecule contracts back to nearly its original bond-length distances. This "breathing" effect is not observed at the second peak bias because resonant tunneling in that case is mainly due to bonding π orbitals. Further details on current-induced forces can be found elsewhere. (*15*)

Ligands and temperature effects

Chen et al. recently reported I-V characteristics of molecules consisting of chains of three benzene rings with ligands replacing various hydrogen atoms in the central ring. (*3-4*) The most interesting result is the large negative differential resistance evidenced by a relatively sharp spike in the I-V characteristic. The spike is found to broaden and shift on the voltage axis with increasing temperature. (*3-4*) The shift, by about 1 V, is very unusual. In semiconductor nanostructures resonant peaks have been found to broaden (by standard electron-phonon interactions) but they never shift appreciably. It has been suggested (*16*) that this effect is due to charging and consequent rotation of the middle ring of the three-ring molecule. However, we note that charging of the molecule would cause additional effects that are not actually observed in the experiments like, e.g., Coulomb blockade. Moreover, we found, by means of ground-state density-

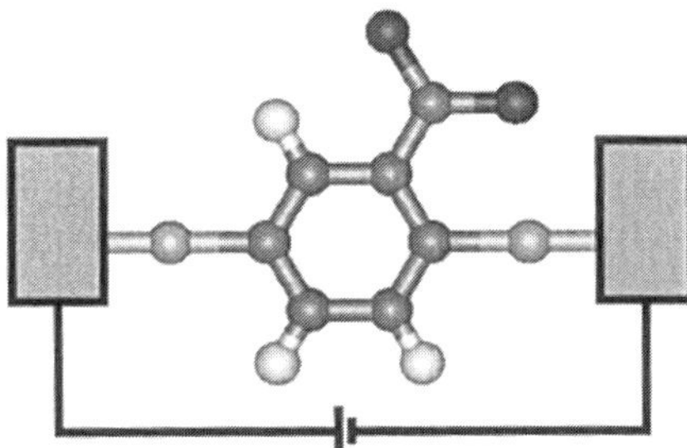

Figure 7: Scheme of the molecular structure investigated. The structure is the benzene-1,4-dithiolate molecule where a H atom is replaced by a NO_2 group. All atoms lie on the plane defined by the carbon ring. The sulfurs attach to ideal metallic leads.

functional calculations on the three-ring molecule, that rotation of the middle ring produces no perceptible change in the electronic structure of the molecule in the vicinity of the highest-occupied molecular orbital (HOMO)-LUMO states, whereas rotation of the ligand does. We thus suggest a different mechanism for the observed phenomena. Since the large negative differential resistance and consequent temperature effect are observed only when an NO_2 ligand replaces one of the hydrogen atoms of the central benzene ring, we pursued the question by calculating the transport properties of a single benzene ring with an NO_2 ligand only (see Fig. 7). We found that the energy levels of the ligand move substantially with increasing voltage and push the π levels into the active energy window for electronic ransport. Therefore the main peak in the current arises primarily from π electrons instead of π^* electrons. We then explored the effect of rotating the ligand. We found that a rotation by 90° shifts the peak to lower voltage by almost 1 V (see Fig. 8), in agreement with the observations. The interpretation is that higher temperatures excite the rotational modes of the ligand. Calculations of the total energy of the molecule as a function of ligand rotation show that the effective rotational quantum of energy is only 3 meV. (*17*) Thus the ligand group can easily rotate at room temperature, practically as a classical rigid rotator and the system will spend most of the time at the highest degree of rotation possible at a given temperature. Moreover, the thermal average of the other vibrational states decays exponentially with the energy distance from the highest occupied vibrational state. The I-V characteristic of the rotating molecule will thus resemble the one depicted in Fig. 8 for the 90° of rotation. Ligand rotation is of course a unique phenomenon of the molecular world, explaining why large voltage shifts are not observed in semiconductor nanostructures.

228

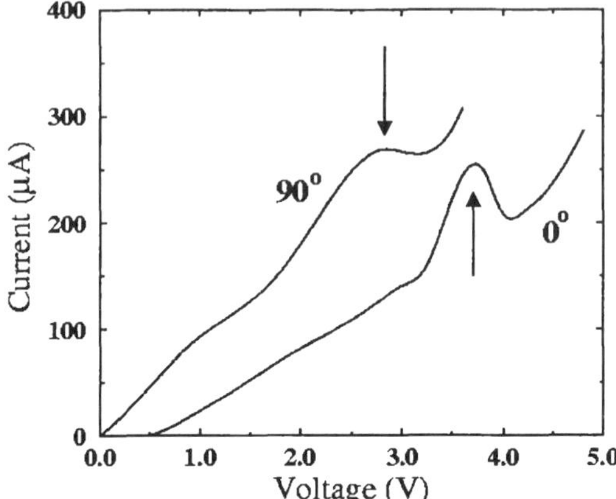

Figure 8: Theoretical I-V curve of the benzene-1,4-dithiolate molecule with a ligand substitution for zero and 90 degrees of rotation of the ligand group with respect to the carbon-ring plane. The vertical arrows indicate the resonant-tunneling maxima for the two configurations.

Conclusions

The calculations summarized above show that theory has now advanced to the point where quantitative predictions can be made about transport in single molecules. Such calculations are expected to play a major role in the development of molecular electronics, the way that simple drift-diffusion calculations of current in semiconductor structures have played in the evolution of silicon-based microelectronics.

Acknowledgements: This work was supported in part by a grant from DARPA/ONR No. N00014-99-1-0351, National Science Foundation Grant No. DMR-98-03768, and by the William A. and Nancy F. McMinn Endowment at Vanderbilt University. Some of the calculations have been performed on the IBM SP2 at the National Partnership for Advanced Computational Infrastructure at the University of Michigan.

1. Aviram A.; Ratner, M.A. *Chem. Phys. Lett.* **1974,** *29,* 277-282.
2. Reed, M.A.; Zhou, C.; Muller, C.J.; Burgin, T.P.; Tour, J.M. *Science* **1997,** *278,* 252-254.
3. Chen, J.; Reed, M.A.; Rawlett, A.M.; Tour, J.M. *Science* **1999,** *286,* 1550-1552.
4. Chen, J.; Wang, W.; Reed, M.A.; Rawlett, A.M.; Price, D.W.; Tour, J.M. *Appl. Phys. Lett.* **2000,** *77,* 1224-1226.

5. Datta, S.; Tian, W.; Hong, S.; Reifenberger, R.; Henderson J.I.; Kubiak, C.P. *Phys. Rev. Lett.* **1997**, *79*, 2530-2533.

6. Samanta, M.P.; Tian, W.; Datta, S.; Henderson, J.I. ; Kubiak, C.P. *Phys. Rev. B* **1996** *53*, R7626- R7630.

7. Emberly, E.G.; Kirczenow, G. *Phys. Rev. B* **2000**, *61*, 5740-5750.

8. Yaliraki, S.N.; Roitberg, A.E.; Gonzales, C.; Mujica, V; Ratner, M.A. J. Chem. Phys. **1999**, *111*, 6997- 7002.

9. Delaney, P; Di Ventra, M.; Pantelides, S.T. *Appl. Phys. Lett.* **1999**, *75*, 3787-3789.

10. Lang, N.D. *Phys. Rev. B* **1995**, 52, 5335-5342. Di Ventra M.; Lang, N.D. Phys. Rev. B **2002**, *65*, 045402.

11. Di Ventra M.; Pantelides, S.T. *Phys. Rev. B* **1995**, *61*, 16207-16214.

12. Di Ventra, M; Pantelides, S.T.; Lang, N.D. *Phys. Rev. Lett.* **2000**, *84*, 979-982.

13. Di Ventra, M.; Pantelides, S.T. ; Lang, N.D. *Appl. Phys. Lett.* **2000**, *76*, 3448-3450.

14. Schon, J.H.; Meng, H.; Bao, Z. Science **2001**, *294*, 2138.

15. Di Ventra, M. ; Pantelides, S.T.; Lang, N.D. *Phys. Rev. Lett.* **2002**, *88*, 046801.

16. Seminario, J.; Zacarias, A.G.; Tour, J.M. *J. Am. Chem Soc.* **2000**, *122*, 3015.

17. Di Ventra, M.; Kim, S.G.; Pantelides, S.T.; Lang, N.D. *Phys. Rev. Lett.* **2000**, *86*, 288-291.

Chapter 17

Computational Studies of Molecular Electronic Devices

Sonja B. Braun-Sand, Olaf Wiest[*], and Jaouad El-Bahraoui

Department of Chemistry and Biochemistry, University of Notre Dame,
South Bend, IN 46556–5670

An overview of the recent progress in the study of molecular electronic devices using computational and theoretical methods is presented. Current work on the molecular design of appropriate molecules, including some recent computations on mixed-valence transition metal complexes from our own group, are presented. We will also discuss the level of theory needed to address design questions at different levels of complexity of the device as well as some of the problems and open questions that persist at this point.

Theoretical and computational studies have been an important aspect of molecular electronics from the very beginning. Some influential concepts in molecular electronics such as the molecular rectifier (*1*) were derived from theoretical considerations rather than from experimental observations. The idea of using molecules as electronic devices continued to inspire a number of different concepts, which were then investigated experimentally.(*2*) It is therefore not surprising that theoretical analysis and computational studies contribute to the recent interest in molecular electronic devices.

As part of the multi-day symposium on molecular electronic devices during the 221st ACS National Meeting in San Diego, CA, two sessions were devoted to computational studies in this field. While the contributions of two invited speakers are highlighted elsewhere in this book, space considerations forbid a complete coverage of the wide-ranging contributions. Here, we will point out some of the recent trends and approaches as discussed in the field as well as some of the open questions that are currently addressed. Some of the issues will be illustrated with examples from the literature and our own recent work.

Generally, computational chemistry can support the development of molecular electronic devices in many aspects. These include both the development of new concepts as well as the interaction with experiment through analysis of observations and quantitatively accurate predictions of properties. While, at least in principle, computational methods are able to address questions at virtually all possible size scales, it is useful to organize the studies according to the questions addressed. In approximate hierarchy of complexity and size of the model systems, these are:

- Computer aided molecular design (CAMD) of compounds with the desired properties
- Analysis of the interface between the molecules and surfaces
- Tools for the design of more complex devices

It is clear from the contributions in this book that the term "molecular electronic devices" covers a wide range of possibilities from molecular computing through sensors to photovoltaic devices and has implications in many other areas.(*3*) Often, computational studies are an integral part of the investigations. In this contribution, as in the sessions during the symposium, the discussion will be limited on an overview of approaches in molecular computing. We will mostly focus on the use of CAMD to molecules with the desired properties. The scope of this contribution is to illustrate the approaches from a molecular point of view using examples taken from the literature and some of our own recent work rather than to provide a comprehensive review of the work in the area. Experimental and theoretical results on the molecule-surface interface are described elsewhere in this book. Finally, we will briefly discuss the applications to the design of large devices.

232

Computational Methodology

The calculation of molecular electronic devices is a daunting task for electronic structure methods. One the one hand, the study of very large systems, involving sometimes hundreds of atoms and possibly thousands of basis functions, is necessary to capture the bulk properties of the materials used. At the same time, a high degree of accuracy is desirable in order to make the quantitative predictions on a given system that give support to the experimental studies. Finally, it is often necessary to consider the dynamical behavior of the molecules. It is clear that all of these requirements will rarely be fulfilled simultaneously and that a careful choice of the model system and the level of theory used will be necessary.

The development of density functional theory (DFT) represents considerable progress towards the calculation of very large systems. Simple functionals like the X_α method were very successful in solid state physics and materials science. However, the accuracy of these methods for molecular systems is often not sufficient to make quantitatively accurate predictions. This is illustrated in Figure 1. The tricobalt nonacarbonyl clusters **1** are stable in their neutral and reduced form.(*4*) They were investigated by our group for potential use as redox active units in molecular electronics using the self-assembled cluster of clusters **2**. Figure 1 shows the deviation of relevant geometrical parameters from the observed values for **1**. All calculations were performed with the 3-21G** basis set using GAUSSIAN 98.(*5,6*)

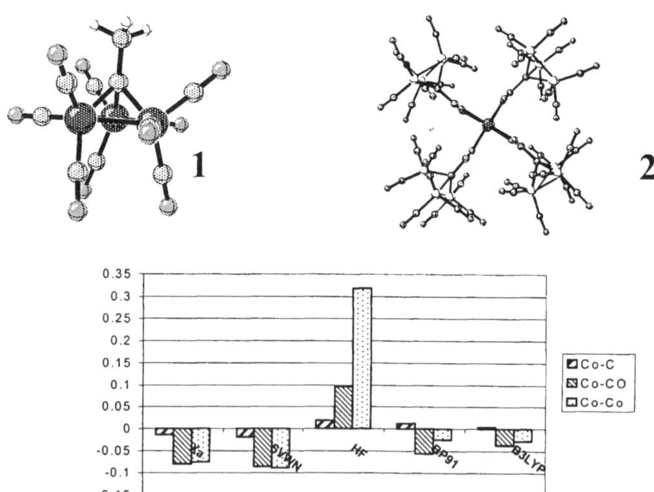

Figure 1: Bond lengths deviation from experimental values of different methods for tricobalt nonacarbonyl complex 1 in Å

It can be seen that Hartree-Fock methods give an unacceptable deviation from the experimentally observed values, (*4*) especially for the metal-metal bonds. This is due to an inadequate treatment of electron correlation in HF calculations. Local DFT methods such as X_α or SVWN perform better, but consistently underestimate the bond lengths as compared to experiment. Gradient corrected or hybrid DFT methods perform significantly better and the results are in very good agreement with experiment. Typical deviations from the experimental values are on the order of a few hundredth of an Angstrom for these systems. We therefore decided to use the B3LYP methodology in our own studies. It should, however, be mentioned that gradient corrected pure DFT methods such as the BP91 functional in combination with fast multipole expansions, which lead to linear scaling of the calculations, might offer an attractive alternative for the study of very large systems. The efficiency advantage offered by these approaches for very large molecular systems can in some cases outweigh the decreased accuracy and the bias towards delocalized wavefunctions.(*7*)

Another problem with the application of DFT in molecular electronics is that the analysis of the results often relies on molecular orbital considerations such as HOMO/LUMO energies and ionization potentials derived from them. The relevance of the Kohn-Sham orbitals derived from DFT calculations is, however, not well established.(*8*) It is for example well known that the HOMO-LUMO gaps calculated by DFT methods are consistently too small. Therefore, the physical meaning of the results derived from such an analysis is not clear.

Finally, it should be mentioned that although most molecular electronic devices rely on a redox reaction for performing their function and therefore have to be studied using electronic structure methods, some devices that rely on large-scale conformational changes have been suggested.(*9*) It has been shown that simple molecular mechanics models are sufficient for the accurate study of such devices.(*10*)

CAMD of Molecular Electronic Devices

Choosing the molecule with the best electronic properties can be very difficult. Given the effort necessary to synthesize and evaluate these fairly complex molecules, the experimental screening of a large number of potential candidates is not efficient. Modeling plays an important role in understanding and predicting these properties, which aids in design of more appropriate molecules with better properties. Several types of molecules have been proposed as possible molecular electronic devices and investigated using computational chemistry.

Molecules in Crossbar Architectures: Molecules that have been proposed are phenylene ethynylene derivatives.(*11*) These are proposed to act as molecular resonant tunneling diodes (RTD) in a crossbar architecture.(*12*) These molecules were shown to exhibit negative differential resistance (NDR) at low temperatures. To try to explain this unusual electrical behavior, theoretical calculations were performed (*13*). In the experimental studies, the system was prepared placing a Gold contact on a Si surface, then attaching the thiols to the surface. Following this, 200 nm of Gold was evaporated onto the monolayer to form the bottom Gold electrode. This is shown schematically in Figure 2. One of the challenges in calculating this system is to adequately represent the molecule as well as the surface contacts.

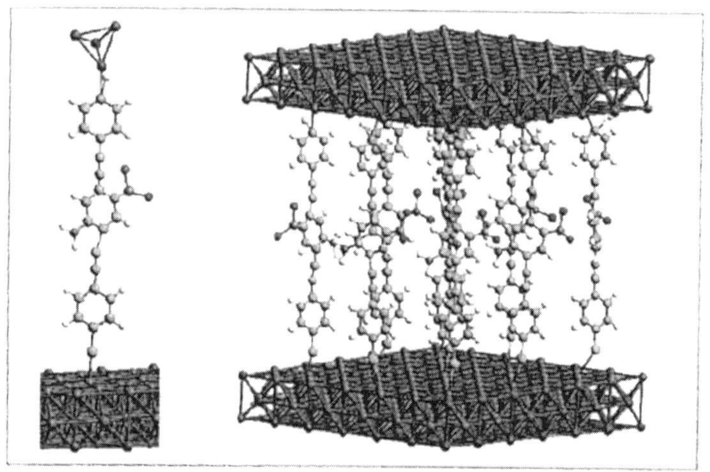

Figure 2. Nanopores containing one (left) and several (right) molecular resonant tunneling diodes on an Gold surface. The molecules are connected to the bottom surface by a sulfur atom, and the top surface is created by vapor deposition. (Reproduced from reference 13.)

When doing the modeling studies, density functional and Green function theories were used in a self-consistent manner. Specifically, B3PW91/6-31G* was used to calculate the neutral molecules and charged species, without the Gold atoms or S linkers.(*12*) When the Gold atoms were included in the calculations, B3PW91 with the LANL2DZ basis set was used for Gold. To model electrical transport in the molecules attached to two metallic contacts, they used a combination of density functional and Green function theories.(*14*) Each contact is modeled by a microscopic and a macroscopic part. The microscopic part consists of three Gold atoms connected to the sulfur end of the molecule, and up to four Gold atoms at the other end. The macroscopic part is located after these Gold atoms and modeled with the scattering Green function for an isolated semi-infinite media. This model allows the calculation of the current-voltage characteristics of the junction, the molecular orbitals participating in electron-transfer, as well as many other properties. Analyzing the frontier molecular orbitals of the molecules in their neutral state gave a qualitative prediction of the ability of the molecule to act as a conductor. (Calculated without sulfur linker or gold atoms.) It was rationalized that a conducting molecular orbital (MO) is fully delocalized along the two metallic contacts, while a nonconducting MO is localized between the metallic contacts. For the molecule that displays NDR (derivative with NH_2 and NO_2 groups on the central phenylene ring) they assumed conduction takes place through the lowest unoccupied molecular orbital (LUMO). It was found with DFT calculations that the NH_2 group localizes the highest occupied molecular orbital (HOMO), and the NO_2 group localizes the LUMO. Neither the HOMO nor the LUMO is delocalized in the neutral state, so it should be nonconducting. In the negatively charged state the LUMO is delocalized, so the molecule should be conducting. Aside from this qualitative picture, by using the combination of density functional theory and Green functions, it was possible to quantify these predictions by calculating the current at each applied external voltage. Theoretical I/V curves for both the neutral and anionic states were plotted. The slope for the anionic state is greater, signifying larger conductance than the neutral state.

Theoretical calculations have also been done on similar model systems for molecular computing. Pantelides examined (*15*) molecules proposed by Reed and coworkers (*16*) using DFT methods. In this example, a single benzene ring is connected to the macroscopic electrode. They were able to calculate the current flow through benzene-1,4-thiolate between two planar metallic electrodes and to reproduce the experimentally observed I/V curves quite accurately. They found the key parameters to be the electronic structure of the organic molecule and the overlap of the wavefunction with the gold atoms representing the surface, whereas the exact geometry of the organic bridge was found to be less important. These results are discussed in more detail in the contribution of DiVentra to this book as well as in several recent publications.(*17*)

Quantum-Dot Cellular Automata: A non-classical approach to information processing, Quantum-dot Cellular Automata (QCA), has been suggested.(*18*) In contrast to the other schemes that have been discussed above, it is not based on a crossbar architecture. The basic concept of QCA is illustrated in Figure 3. Four particles that can be reversibly charged are arranged in a square planar fashion. When two additional charges are added, the "extra" charges occupy opposite corners of the square due to Coulombic repulsion. This results in two energetically degenerate but distinguishable quantum states that could correspond to a 1 and 0 in binary code, as shown in Figure 3. Several

236

logical devices based on QCA have been demonstrated using metal nanoparticles.(*19*) However, these relatively large particles require the use of low temperatures to run the experiments due to the close energy spacing of the quantum levels. High-temperature QCA devices could be achieved using molecular implementations of the QCA paradigm.

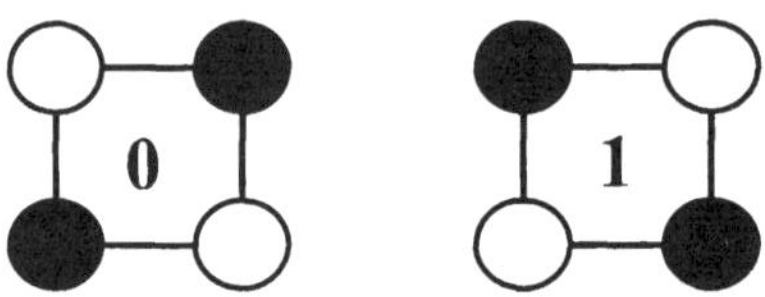

Figure 3: Quantum states encoding information in QCA

Mixed valence compounds can be envisioned as molecular QCA cells. Although the use of mixed valence complexes in molecular electronics has been proposed previously,(*20*) their use in molecular QCA differs significantly from these earlier concepts. Rather than current transfer through molecules, as required in molecular rectifier and crossbar architectures, it focuses on electron tunneling within a square molecule, encoding the information in different charge distributions in a molecule. In addition to the cluster complexes **2**, molecular square analogues of the well-studied Creutz-Taube complex (*21*) and similar complexes (*22*) with various bridging ligands have been suggested as molecular QCA cells. The crucial properties for the degree of electron localization in the molecular QCA cell are the electronic communication between the redox active centers as expressed by the electronic coupling matrix element, the molecular geometry (especially the distance between the redox active metal centers), and the response of the electron distribution to an external field. The electronic and geometric properties can be varied by changing the bridging ligand. The bridging ligand should provide a barrier to electron transfer between the metal centers, so that random switching does not occur at room temperature.

Several of these questions can be addressed using computational methods. It is hoped that calculation of these and similar complexes will reproduce their structural and electronic properties and explain their molecular origin. This methodology can then be used to predict the properties of other complexes that have not yet been synthesized. Some of the properties of interest and that can be calculated include how localized or delocalized the electrons in the complex are, the dipole moment in the ground state, and how the dipole moment changes as the system is biased by an electric field. It should be possible to use calculations such as these to design a complex with the desired properties. Results from this approach are shown in Figure 4. Two molecular two-dot QCA cells, the Creutz-Taube ion **3** and a recently synthesized analog **4** are shown in the top part. A square, four-dot molecular QCA cell **5,** which has not yet been prepared, is shown in the lower part of Figure 4.

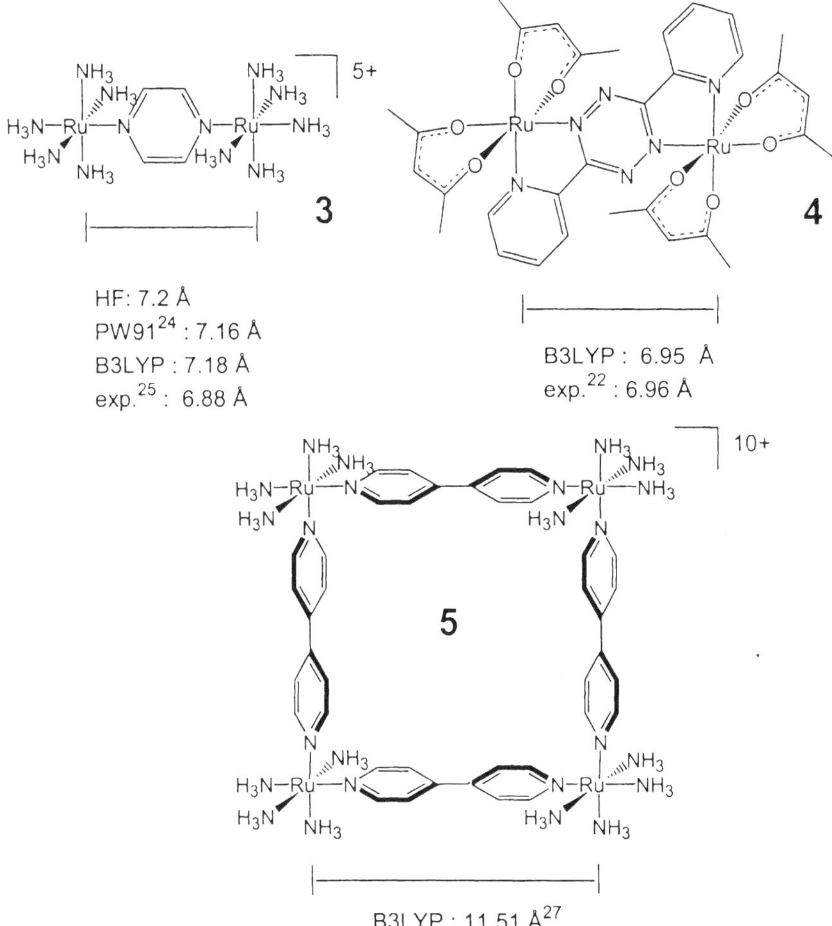

Figure 4. Calculated and experimental geometries for two-dot and four-dot molecular QCA cells

The QCA concept has several advantages from the computational point of view. There is no need to include micro- or macroscopic metal electrodes in the calculations because no current is required for these molecules to function. All processes associated with information processing occur in the ground state. Finally, there are no large-scale conformational changes associated with the intramolecular electron transfer. On the downside, molecular QCA cells are by definition two-state systems and a rigorous treatment would require a multiconfiguration wavefunction which is computationally too costly for any realistic molecule. However, we (23) and others (6a,d) have shown in the past that DFT methods yield reasonable results for such systems.

Geometry optimizations on the complexes were done using density functional and Hartree-Fock theories, in particular B3LYP/3-21G and HF/3-21G.(24) As can be seen from the results in Figure 4, the Ru-Ru distance in **3** is significantly overestimated as compared to the experimental value.(25) This is due to the electrostatic repulsion of the positive charges on the metal atoms in these gas phase calculations and is expected to be a general problem in the calculation of highly charged complexes. Consequently, the calculated Ru-Ru bond length of 6.95 Å in the neutral complex **4** is in excellent agreement with the experimental value if 6.96 Å from x-ray crystallography. This gives reasonable confidence into the reliability of the calculated geometry of the mixed-valence complex **4**$^+$ with a Ru-Ru distance of 6.76 Å. Although the structure of **4**$^+$ has not yet been determined, the calculated delocalized structure is in good agreement with the experimental characterization of **4**$^+$ as a Robin-Day class III complex.(26,)

These calculations can then be extended to elucidate the geometric and electronic structure of more complex QCA molecules such as the molecular four-dot cell **5**, which has not yet been synthesized. Based on studies of the corresponding two-dot cell with a 4,4'-bipyridine bridge, this four-dot cell should have Robin-Day class II characteristics and a Ru-Ru distance of 11.51 Å. The computational classification of compounds according to the Robin-Day scheme before they are synthesized is useful because of the inherent bistability of Robin-Day class II molecules. They can also support the interpretation of experimental measurements, which do not always allow an unambiguous assignment of the degree of localization in mixed-valence complexes.

The analysis of the electronic structure of these and similar compounds then gives valuable insights into the origin of the observed redox behavior. The degree of electron localization on one of the metal centers, and thus the degree of bistability of the system, can for example be correlated to the energy gap between the SOMO and the bridging ligand MOs (assuming electron transfer occurs by going from one metal atom to the bridging ligand to the other metal atom). As expected, the less conjugated the bridging ligand is, the larger the energy gap between the orbitals. This should correspond to more electron localization on the metal centers.(28) The degree of electronic coupling can also be calculated quantitatively using a variety of approaches, as discussed by M. Newton in this book.

Modeling of Complex Devices

Calculations of systems such as the ones described above represent the size limit of current high accuracy molecular electronic structure methods. The extension to larger devices consisting of multiple molecules from first principles will only be possible at much lower levels of approximation and thus give only qualitative insights. The simulation of entire circuits will be out of reach even for simple *ab initio* quantum mechanical methods at least for some time. A possible solution that can speed up the simulation by several orders of magnitude is the use of parameterized models. To achieve the necessary accuracy, the parameters can be fitted to reproduce the results from high-level calculations of smaller building blocks as described above. These can then be represented either in semiempirical approaches or, even faster, as local charges in an electrostatic multipole approach. These methods are fast enough to describe for example the dynamic behavior of the systems. However, their relatively low accuracy typically allows only qualitative predictions.

While many tight-binding calculations with adjustable parameters belong in this category, some of the most advanced and useful applications involve the development of design tools for molecular circuits.(*29*) The approach used is often similar to the functionally related SPICE simulations. However, many of the parameters required for the simulation such as FMO energies, and partial charges are not directly accessible and are therefore extracted from quantum mechanical calculations rather than experimental measurements.

Conclusions and Outlook

Computational chemistry and conceptual theoretical chemistry has played, and continues to play, an important role in all areas of development of molecular electronic devices. With the development of new computational methods, particularly DFT calculations, and the rapid increase in CPU performance, it is now possible to move from conceptual considerations of molecular electronics to quantitatively accurate predictions of the geometric and electronic structure of the molecules involved. In addition, initial steps have been taken to model more complex molecular assemblies and the interface to the macroscopic world using first principles methods.

Similar to organization of this overview, the challenges facing the field in the next few years can be grouped according to the complexity of the systems to be considered:

- For the design and screening of molecules in CAMD, the accuracy and throughput of molecules needs to be increased. This is especially true for systems where redox reactions are coupled to large-scale conformational changes, *(8,30)* which are currently not easily accessible using computational methods. This situation could potentially be improved using multi-layered methods such as QM/MM or ONIOM.(*6*)

- The investigation of the interface between bulk materials and the individual molecules (in other words, between the macroscopic and the microscopic world) is likely to require new methods that merge the accuracy of molecular electronic structure theory with the conceptual frameworks of solid state calculations.

- The simulation of large devices and dynamical behavior within these devices is still in its developmental stages. Future developments will have to improve the representation of the molecular structure in these parameterized models in order to achieve higher accuracy.

Acknowledgements

We gratefully acknowledge financial support of this research by the Research Corporation (Grant RI0117) and the DARPA/ONR Moletronics program (Grant N00014-99-1-0472) and generous allocation of computer resources by the Notre Dame Office of Information Technology and the National Center of Supercomputer Applications at UIUC (Grant CHE90004 to J. E.-B.). S. B. Braun-Sand is a recipient of a NSF Graduate Fellowship.

References

1   Aviram, A.; Ratner, M. A. *Chem. Phys. Lett.* **1974**, *29*, 277.

2   For an earlier overview of theoretical concepts in molecular electronics, compare for example: Munn, R. W. In: *Molecular Electronics* Ed.: Lazarev, P.I; Kluwer, Dordrecht 1991 pp. 1-8

3   For an overview, compare e.g.: a) Mirkin, C. A.; Ratner, M. A. *Ann. Rev. Phys. Chem.* **1992**, *43*, 719-754.

4   Sutton, P. W.; Dahl, L. F. *J. Am. Chem. Soc.* **1967**, *89*, 261-268. b) D'Agostino, M. F.; Frampton, C. S.; McGlinchey, M. J. *Organometallics* **1991**, *10*, 1383-1390.

5   We also investigated other double-zeta quality basis sets, which gave very similar results.

6   Gaussian 98, Revision A.9, M. J. Frisch, G. W. Trucks, H. B. Schlegel, G. E. Scuseria, M. A. Robb, J. R. Cheeseman, V. G. Zakrzewski, J. A. Montgomery, Jr., R. E. Stratmann, J. C. Burant, S. Dapprich, J. M. Millam, A. D. Daniels, K. N. Kudin, M. C. Strain, O. Farkas, J. Tomasi, V. Barone, M. Cossi, R. Cammi, B. Mennucci, C.

Pomelli, C. Adamo, S. Clifford, J. Ochterski, G. A. Petersson, P. Y. Ayala, Q. Cui, K. Morokuma, D. K. Malick, A. D. Rabuck, K. Raghavachari, J. B. Foresman, J. Cioslowski, J. V. Ortiz, B. B. Stefanov, G. Liu, A. Liashenko, P. Piskorz, I. Komaromi, R. Gomperts, R. L. Martin, D. J. Fox, T. Keith, M. A. Al-Laham, C. Y. Peng, A. Nanayakkara, C. Gonzalez, M. Challacombe, P. M. W. Gill, B. Johnson, W. Chen, M. W. Wong, J. L. Andres, C. Gonzalez, M. Head-Gordon, E. S. Replogle, and J. A. Pople, Gaussian, Inc., Pittsburgh PA, 2000.

7   This bias is a general problem of DFT based methods due to a non-exact treatment of self-interaction and can lead to qualitatively wrong results in some cases: a) Sodupe, M.; Bertran, J.; Rodriguez-Santiago, L.; Baerends, E. J. *J. Phys. Chem. A* **1999**, *103*, 166-170. b) Braïda, B.; Hiberty, P. C. *J. Phys. Chem. A* **1998**, *102*, 7872-7877. c) Bally, T.; Sastry, G. N. *J. Phys. Chem. A* **1997**, *101*, 7923-7925. d) Oxgaard, J. Wiest, O. submitted for publication.

8   For a more detailed discussion of this, compare for example: a) Koch, W.; Holthausen M. C. *A Chemists Guide to Density Functional Theory.* Wiley-VCH, New York, 2000, p.47-52 and references cited therein. b) Stowasser, R., Hoffmann, R. *J. Am. Chem. Soc.* **1999**, *121*, 3414-3420. C) Baerends, J. E. *Theor. Chem. Acc.* **2000**, *103*, 265-269. d) Bickelhaupt, F.M., Baerends, E.J. *Rev. Comp. Chem.* **2000**, *15,* 1-86

9   Carter, F. L. In: *Molecular Electronic Devices Vol. 1*, F. L. Carter, Ed. Marcel Dekker, New York 1982, p.51-72.

10 Rueckes, T.; Kim, K.; Joselevich, E.; Tseng, G.Y.; Cheung, C.L.; Lieber, C.M. *Science* **2000**, *289*, 94-97.

11 Chen, J.; Reed, M. A.; Rawlett, A. M.; Tour, J. M. *Science*, **1999**, *286*, 1550-1552.

12 Seminario, J. M.; Zacarias, A. G.; Derosa, P. A. *J. Phys. Chem. A*, **2001**, *105*, 791-795.

13 Seminario, J. M.; Zacarias, A. G.; Tour, J. M. *J. Am. Chem. Soc.*, **2000**, *122*, 3015-3020.

14 Derosa, P. A.; Seminario, J. M.; *J. Phys. Chem. B*, **2001**, *105*, 471-481.

15 Pantelides, S. T.; Di Ventra, M.; Lang, N. D. *Physica B*, **2001**, 296, 72-77.

16 Reed, M. A.; Zhou, C.; Muller, C. J.; Burgin, T. P.; Tour, J. M. *Science*, **1997**, *278*, 252-254.

17 Lang, N. D. *Phys. Rev. B.*, **1995**, *52*, 5335-5342.

18 Tougaw, P. D.; Lent, C. S. *J. Appl. Phys.* **1994**, *75*, 1818-1825.

19 a) Orlov, A. O.; Amlani, I.; Bernstein, G. H.; Lent, C. S.; Snider, G. L. *Science*, **1997**, *277*, 928-930. b) Amlani, I.; Orlov, A.O.; Toth, G.; Bernstein, G.H.; Lent, C.S.; Snider, G.L. *Science*, **1999,** *284*, 289-291. c) Orlov, A.O.; Kummamuru, R.K.; Ramasubramaniam, R.; Toth, G.;

Lent, C.S.; Bernstein, G.H.; Snider, G.L. *Appl. Phys. Lett.* **2001**, *78*, 1625-1627.

20 a) Launay, J.-P. In: *Molecular Electronic Devices Vol. 2*, F. L. Carter, Ed. Marcel Dekker, New York 1987, p.39-55. b) Woitellier, S.; Launay, J. P.; Joachim, C. *Chem. Phys.* **1989**, *131*, 481-488. c) Joachim, C.; Launay, J. P. *Chem. Phys.* **1986**, *109*, 93-99. d) Joachim, C.; Launay, J. P. *Nov. J. Chim.* **1984**, *8*, 723-728.

21 Creutz, C. *Progr. Inorg. Chem.* **1988**, *30*, 1-73.

22 Chellamma, S.; Lieberman, M. *Inorg. Chem.* in press.

23 Saettel, N. J.; Oxgaard, J.; Wiest, O. *Europ. J. Org. Chem* **2001**, *6*, 1429-1439.

24 For other computational studies of **3**, compare: Bencini, A.; Ciofini, I.; Daul, C. A.; Ferretti, A. *J. Am. Chem. Soc.* **1999**, *121*, 11418-11424 and references cited therein.

25 Fürholz, U.; Bürgi, H.-B.; Wagner, F. E.; Stebler, A.; Ammeter, J. H.; Krausz, E.; Clark, R. J. H.; Stead, M. J.; Ludi, A. *J. Am. Chem. Soc.* **1984**, *106*, 121-123.

26 Robin, M. B. Day, P. *Adv. Inorg. Chem. Radiochem.*, **1967**, *10*, 247

27 Based on the calculated geometry of the 4,4'-bipyridyl bridged di(pentamino) ruthenium complex.

28 Braun-Sand, S. B.; Wiest, O. to be submitted.

29 E.g.: Ellenbogen, J.C.; Love, J.C. *Proc. IEEE* **2000**, *88*, 386-426. b) Ami S, Joachim C. *Nanotechnology* **2001**, *12*, 44-52. c) Mohan, S.; Sun, J. P.; Mazunder, P.; Haddad, P.; Haddad, G. I. *IEEE Transactions on Computer-Aided Design of Integrated Circuits and Systems* **1995**, *14*, 653-662.

30 For an advanced systems based on this concept, see: Collier CP, Mattersteig G, Wong EW, Luo Y, Beverly K, Sampaio J, Raymo FM, Stoddart JF, Heath JR *Science* **2000**, *289*, 1172-1175 and references cited therein.

Chapter 18

Excitonic Solar Cells: The Physics and Chemistry of Organic-Based Photovoltaics

Brian A. Gregg

National Renewable Energy Laboratory, 1617 Cole Boulevard, Golden, CO 80401

The physical principles underlying the photovoltaic effect in organic-based solar cells are described and compared to those in conventional solar cells. The photo induced chemical potential gradient created by the interfacial dissociation of excitons is an important, and frequently overlooked, driving force for charge separation in organic photovoltaic cells. The photovoltage is proportional to the sum of the chemical potential gradient and the electrical potential gradient (band bending or built-in potential). Photovoltages up to ~1 V have been obtained in cells without any built-in potential, despite misconceptions that this is impossible. The importance of the photo generated interfacial electric field is emphasized, and the difficulties in disentangling exciton transport length measurements from interfacial quenching rates are discussed. The advantages and disadvantages of increasing the exciton-dissociating surface area by structuring the substrates, or by structuring the films on planar substrates, are considered.

Introduction

Interest in non-conventional solar cells has been increasing rapidly since the discovery of the high-efficiency dye-sensitized solar cell, DSSC, by Grätzel and coworkers in the late 1980s (1,2). More recently, substantial advances in conversion efficiency also have been achieved with photovoltaic (PV) cells based on molecular semiconductors (3-5) and on conducting polymers (6-9). A number of variations on these three themes have been described and more are expected, yet they all differ in some fundamental respects from the conventional solar cells epitomized by silicon p–n junction devices. The purpose of this article is to briefly review the basic principles common to all known solar cells and to show how different aspects of these common principles are accentuated in the different types of cells. The primary difference between conventional PV cells and organic PV (OPV) cells (among which I include DSSCs) is that light absorption results in the formation of electrically neutral excited states in OPV cells in place of the free electron–hole pairs formed in conventional PV cells. The dissociation of these excited states at an interface can produce a photovoltaic effect in the absence of band bending, and this is probably the most misunderstood aspect of OPV cells. However, the same basic equations that describe conventional PV cells also show that a PV effect is possible without band bending. The importance of understanding the role of the photo induced chemical potential gradient in OPV cells is stressed.

Fluxes and Forces

The current density equation (10) describes the flux of charge carriers through semiconductor devices. Under the conditions most relevant to solar cells (constant temperature and pressure, no magnetic fields), the current density, J, is determined by the electric field, E, and the chemical potential gradient ∇n. Only the equations for electrons are shown here.

$$J_n = n\,\mu_n\,q\,E + q\,D_n\,\nabla n \qquad (1)$$

where n is the concentration of electrons, μ_n and D_n the mobility and diffusion coefficient, respectively, and q is the electrical charge. The total current through the cell is $J = J_n + J_p$ where J_p is the current density of holes. Equation 1 is valid both in the dark and under illumination. By recognizing that the gradient of the conduction (or valence) band energy, known as the band bending, is a measure of the electric field, $\nabla E_{cb} = q\,E,$ we can rewrite eq. 1 in a more symmetrical form in terms of the sum of two potential gradients:

$$J_n = n\,\mu_n\,\nabla E_{cb} + q\,D_n\,\nabla n \tag{2}$$

The flux of electrons is thus proportional to the sum of two forces: the gradient of the electrical potential, ∇E_{cb}, and the gradient of the chemical potential, ∇n. ∇n is also called the concentration gradient of electrons, but it is essential to realize that it constitutes a force acting on electrons that is equivalent to ∇E_{cb}, the electrical force. The chemical potential gradient is not important in metals because they have such a high density of electrons and positive lattice sites that no substantial concentration gradient can form. But ∇n becomes important in materials such as semiconductors where large concentration gradients exist, for example, at a conventional p–n junction or at an exciton-dissociating interface in an OPV cell. A band diagram of a conventional p–n homojunction solar cell at equilibrium (left side of Figure 1) shows the energies of the conduction and valence band edges, E_{cb} and E_{vb}, and the band bending that occurs at the junction.

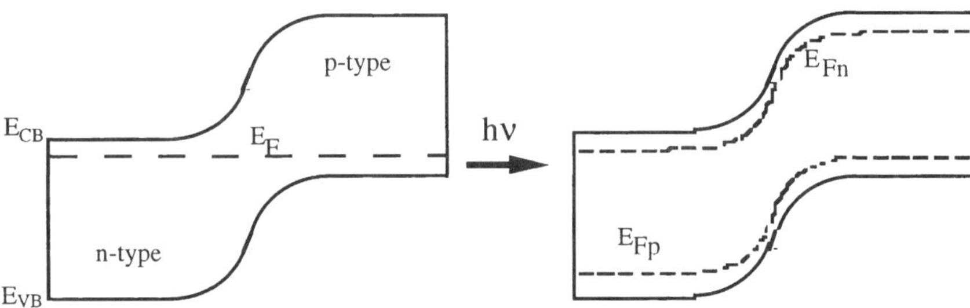

Figure 1. *A schematic band diagram of a conventional p–n homojunction solar cell at equilibrium on the left and at short circuit under* uniform *illumination on the right. The energies of the conduction and valence band edges are E_{cb} and E_{vb}, respectively. E_F is the Fermi level at equilibrium while E_{Fn} and E_{Fp} are the quasi Fermi levels of electrons and holes, respectively, under illumination.*

The expression for the open circuit photovoltage, V_{oc}, can vary somewhat depending on the specific mechanisms involved in the photovoltaic effect (10-12) but it usually has the following form:

$$V_{oc} = n_q k\, T/q\; ln[(J_l/J_0) + 1] \qquad (3)$$

where n_q is the diode quality factor, k is Boltzmann's constant, T is the temperature, J_l is the photocurrent and J_0 is the reverse saturation (or exchange) current density in the dark. Equation 3 shows that a photocurrent, generated by whatever mechanism, must always be accompanied by a corresponding open circuit photovoltage which is simply the potential required to counteract the photocurrent and make it go to zero.

The current density equation (eq. 2) shows that even if there is no band bending, that is, even if $\nabla E_{cb} = 0$, a photovoltaic effect can still be achieved if light absorption creates a chemical potential gradient, ∇n. Thus, without assuming anything other than the validity of the current density equation, the misperception that a photovoltaic effect is impossible without band bending can be eliminated. And, turning the argument around, the measured photovoltage in a solar cell cannot be equated necessarily to the magnitude of the band bending, since this ignores the effect of ∇n. These conclusions are crucial to understanding the non-conventional solar cells. They are not widely appreciated, however, probably because the mechanism of charge separation in the well-known p-n homojunction solar cells does, in fact, require that ∇E_{cb} be non-zero. But this is required only by the mechanism for charge separation in these cells, and is not an absolute requirement for a photovoltaic effect.

The electrochemical potential of electrons and holes, that is, the Fermi level (see Figure 1), is constant everywhere in the cell at equilibrium by definition. Upon illumination (right side of Figure 1), the electrons and holes are no longer in equilibrium and thus there are separate electrochemical potentials, quasi Fermi levels, for electrons and holes. The quasi Fermi level for electrons is

$$E_{Fn} = E_{cb} + k\, T\, ln(n/N_c) \qquad (4)$$

where N_c is the density of states at the bottom of the conduction band. The gradient of the quasi Fermi level, ∇E_{Fn}, contains the gradients of both the electrical and chemical potentials and thus represents the thermodynamic driving force for electron motion in illuminated solar cells. Therefore, the current density can be expressed in its simplest form as a function of the gradient of the quasi Fermi level:

$$J_n = n\, \mu_n\; \nabla E_{Fn} \qquad (5)$$

As a solar cell charges from the short circuit condition toward the open circuit voltage, VE_{Fn} and the corresponding quantity for holes, VE_{Fp}, decrease. As the gradients approach zero, that is as the quasi Fermi levels become flat, the driving force for charge transport approaches zero and the cell reaches its open circuit photovoltage, V_{oc}. The maximum possible V_{oc} is therefore set by the maximum photo induced splitting of the quasi Fermi levels. If the number of photogenerated charge carriers is directly proportional to the light intensity, the splitting will increase with the logarithm of the light intensity according to eq. 4, resulting in the logarithmic increase in V_{oc} with light intensity expected from eq. 3. This behavior is expected *in general* for solar cells and is not dependent on the detailed mechanisms of charge carrier generation, separation and recombination. By the same token, one cannot deduce that a solar cell functions by the conventional mechanism just because its J-V behavior is similar to conventional solar cells.

Although calculating the exact quasi Fermi levels generally requires the numerical solution of several coupled differential equations, their qualitative behavior is quite easy to estimate and can be used to describe and compare the different types of solar cells.

Conventional Solar Cells

It is instructive to consider what happens when an n-doped and a p-doped slab of the same semiconductor material are conceptually brought into contact to form a p–n junction. Before contact there is no electrical potential difference between the two sides, but a large chemical potential difference. Upon contact, current flows according to eq. 2: carriers flow until the electrical potential difference created by the loss of majority carriers is exactly equal and opposite to the chemical potential difference between the two sides. This is the equilibrium condition shown on the left side of Figure 1. Light absorption leads to a substantial increase in the minority carrier density (electrons on the p-type side and holes on the n-type side) and thus to a decrease in the original chemical potential difference between the two sides of the cell. If the electrical potential difference is held constant (i.e., maintained at short circuit as shown in the right side of Fig. 1), electrons will flow back to the n-type side and holes back to the p-type side driven by the photo induced difference in chemical potentials. If the circuit is opened, the cell will charge up to V_{oc}, flattening the bands by an amount equal to the difference between the photoinduced and the equilibrium chemical potentials.

It is clear from eq. 2 that a photovoltaic effect can be achieved without any band bending ($VE_{cb} = 0$); yet years of experience with conventional solar cells shows that bandbending is required for a conventional homojunction photovoltaic cell. How do we reconcile this apparent contradiction? What the simple theory of eq. 2 does not contain is a charge separation mechanism, an essential process in solar cells where photo generated electrons must be separated from photo generated holes to prevent the energy-wasting recombination. It is the charge separation mechanism that most clearly distinguishes conventional solar cells from OPV cells. In conventional semiconductors, light absorption leads directly to the production of free electrons and holes throughout the bulk wherever light is absorbed. The only way to efficiently separate free charge carriers from each other in the same material (homojunction cells) is with an electric field. (Separation can be achieved, however, in heterojunction cells with type 2 band offsets without an electric field – this process is related to that which occurs in excitonic solar cells.) When charge separation occurs only via an electric field, the magnitude of the built-in potential sets an upper limit to V_{oc}; not because of thermodynamic necessity, but because of the mechanism for charge separation. This is the case in silicon solar cells; and most other conventional PV cells, even heterojunction cells, also rely on the built-in potential (band bending) to separate carriers. Excitonic solar cells, however, often rely on a different mechanism for charge separation and thus V_{oc} is only partially determined by, or even independent of, the band bending.

Excitonic Solar Cells

Excitons play only a small role in conventional PV cells at room temperature (13), but they have a decisive influence on the behavior of organic-based PV cells. This is probably the most fundamental difference between the two types of PV cells. Excitons come in a number of guises (14) – singlet and triplet excitons, Frenkel and Wannier excitons, charge transfer excitons and exciton-polarons – but for the purpose of this paper these distinctions are unnecessary, and I will use only the most fundamental definition of an exciton: a mobile, electrically neutral, optically excited species. Excitons exist because they do not have enough energy to dissociate into an electron and a hole in the pure bulk phase. The difference between the energy of a free electron-hole pair (the electrical bandgap) and the exciton energy (optical bandgap) is the exciton binding energy; typically 200 - 400 meV. Exciton formation is favored in low dielectric constant material where the Coulombic attraction between electron-hole pairs is large compared to the thermal energy ($kT = 26$ meV at room temperature). The spatial extent of an exciton is often around 1 nm; thus very high electric fields, ~200 - 400 mV/nm = 2 - 4 x 10^6 V/cm, would be required to cause them to dissociate in the bulk material. Such fields are unlikely to occur in most low-doped OPV cells under normal operating conditions. Therefore, the two more important mechanisms for producing charge carriers from excitons involve dissociation at bulk trap sites and dissociation at interfaces. Carriers can be generated throughout the bulk when excitons dissociate at traps, for example, by trapping the hole and leaving the electron in the conduction band (Figure 2a). This process somewhat resembles carrier formation in conventional PV cells and can be described approximately by the same model. However, this mechanism would require a high concentration of traps in order to dissociate most of the excitons, and it results in only one mobile carrier; it is thus unlikely to result in an efficient OPV cell.

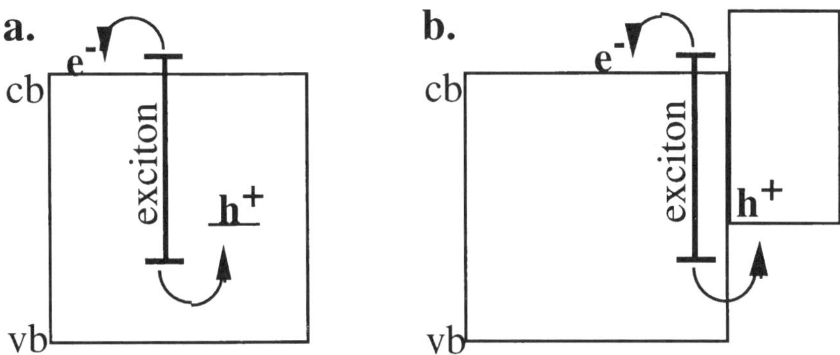

*Figure 2. Schematic diagrams showing **a**: exciton dissociation at a bulk trap state leading to a trapped hole, h^+, and a free electron, e^-; and **b**: exciton dissociation at an interface leading to a free hole in one phase and a free electron in the other.*

The more interesting and potentially more useful mechanism for exciton dissociation occurs at organic/organic interfaces or at organic/inorganic interfaces. When the difference in free energy between an electron on one side of the interface and a hole on the other is less than the exciton energy, dissociation can occur (Figure 2b). This mechanism results in carrier generation and carrier recombination only at the interface, not in the bulk. Interfacial exciton dissociation is apparently the dominant mechanism for photocurrent production in many OPV cells (6, 8, 9, 15-22) and this distinguishes them mechanistically from conventional PV cells. To our knowledge, the consequences of this mechanism for the band structure and for the current-voltage characteristics of an illuminated OPV cell have not yet been described. Thus we provide a qualitative description below and will provide a more quantitative description in later publications.

The first unambiguous example of a photovoltaic effect without any bandbending was reported in a symmetrical cell consisting of a liquid crystal porphyrin sandwiched between two ITO electrodes (15). The effect was explained as resulting from asymmetric exciton dissociation at the illuminated electrode: electron injection from the exciton into the ITO electrode was kinetically more facile than hole injection, leading to an excess concentration of holes in the porphyrin near the ITO interface. This photo-induced chemical potential gradient drove the photovoltaic effect. Photovoltages of up to $V_{oc} \approx 1$ V have been achieved in similar cells with no band bending by derivatizing the electrodes to make the exciton dissociation process more asymmetric (16). In further work, we showed that the chemical potential gradient can overwhelm the built-in electric field in some cases, causing the photocurrent to flow in the direction opposite to that expected from the difference in work functions of the electrodes (17). The PV effect in DSSCs is also driven by interfacial "exciton" dissociation (although the "exciton" in this case does not have to move anywhere), and we showed that the difference in work functions between the two electrodes plays little or no role in determining V_{oc} (22). Also in this case, photovoltages of around 1 V have been achieved without any built-in potential. The DSSC has reached efficiencies of ~10%, proving that it is possible to make commercially viable solar cells without relying on band bending for charge separation. Taken together, these results clearly demonstrate the inadequacy of the conventional solar cell model (V_{oc} is determined by band bending, etc.) to describe OPV cells. The conventional model implicitly assumes a specific mechanism for charge separation that is valid for silicon p-n homojunctions, but not for OPV cells; it neglects the photo induced chemical potential gradient caused by interfacial exciton dissociation, and thus neglects one of the two major forces in OPV cells.

The description of the J-V behavior of solar cells requires the simultaneous solution of eq. 1, the Poisson equation and the continuity equation for electrons:

$$-\frac{1}{q}\frac{dJ_n(x)}{dx} = G(x) - U(x) \qquad (6)$$

where $G(x)$ and $U(x)$ are the carrier generation and recombination rates, respectively. According to eq. 6, the current density is a function of position in conventional solar cells because generation and recombination occur in the bulk semiconductor. This condition is an integral part of the description of conventional PV cells, but it is inappropriate for OPV cells based on interfacial exciton dissociation. In these cells, $G(x) = U(x) = 0$ everywhere except at the exciton-dissociating interface, so the current density is not a function of position. A quantitative model for these cells is under development, but for now, a qualitative description is given of one of the most important aspects: the interfacial electrical potential under illumination.

When excitons dissociate at an interface, they generate a chemical potential gradient that can generate a photocurrent, as discussed above; but they also generate an electrical potential gradient that opposes charge separation due to the Coulomb attraction between the sheets of electrons and holes. In conventional solar cells this leads to a decrease in the built-in voltage or a "flattening" of the bands. The same can occur in OPV cells, but the interfacial electric field can be very high because all carrier generation occurs in a very narrow region of the cell, within ~1 nm of the interface. We predict qualitatively, and the idea is supported by preliminary numerical calculations, that under solar intensities the photogenerated interfacial field can become larger than the built-in field. This may be a major, and heretofore unrecognized, loss mechanism in excitonic OPV cells. The situation is illustrated in Figure 3 where it is assumed that a typical bilayer OPV cell is sandwiched between electrodes of different work functions, resulting in band bending at equilibrium. Because of the low doping density of most organic semiconductors, the field is assumed to drop uniformly across the device in the dark, rather than just near the junction as it does in highly doped inorganic semiconductors. Thus the equilibrium field at the interface may be quite low. Upon illumination, the photogenerated field caused by exciton dissociation will drop mostly across a few molecular layers at the interface. At high light intensity it will likely overwhelm the bulk field, that is, bend the bands in the "wrong" direction, and thus promote interfacial recombination (right side of Figure 3). In this case, the equilibrium band bending is mainly effective for accelerating charge transport through the bulk, but it is ineffective for the crucial process of separating carriers at the interface. The deleterious photoinduced interfacial electric field may be partially responsible for the decrease in fill factor and power conversion efficiency commonly observed in OPV cells at high light intensities.

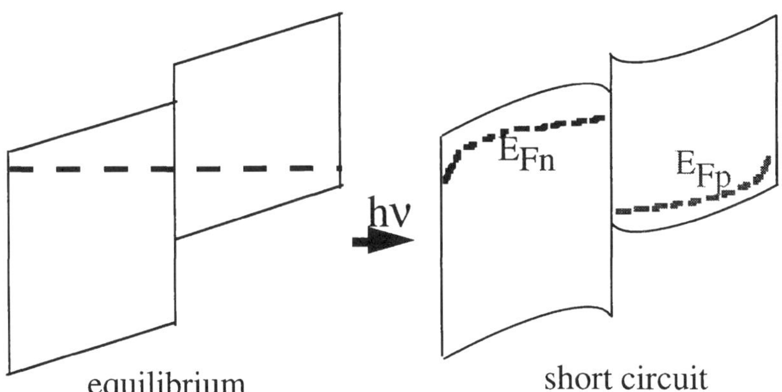

Figure 3. Qualitative band diagrams of an organic heterojunction showing the band offset at the interface and the assumed band bending at equilibrium. Upon illumination, the photo induced field at the interface may overwhelm the equilibrium field and hinder carrier separation at high light intensity.

OPV cells like the one shown schematically in Fig. 3 are commonly referred to as "p-n junction" cells. This terminology is misleading in several respects and tends to obscure more than it explains about OPV cells. For example, the doping density in most organic semiconductors is very low and the dopants are generally adventitious – neither their identity nor their concentration is known or controlled. And known dopants like oxygen are usually free to diffuse through the film, ensuring that no dimensionally or electrically stable junction can be formed. Furthermore, the essential characteristic of an OPV junction is that excitons dissociate there; a mechanistically distinct process from electron-hole pair separation at an electrical p-n junction. Finally, real p-n junction solar cells are minority carrier devices: their efficiency depends on the ability of, say, photo generated electrons to escape from the p-type side without recombining. But excitonic heterojunctions, the usual OPV cells, are majority carrier devices: only electrons are present in one phase and only holes in the other; recombination occurs at the interface between the two, and not in the bulk. Thus there are a number of fundamental mechanistic differences between conventional p-n junction solar cells and OPV heterojunction solar cells. The fact that their J-V characteristics can be described by the same equations is expected, since all solar cells should show qualitatively similar behavior, as described above. But to correctly interpret this behavior requires an understanding of the kinetic mechanism that governs the PV effect. Using the conventional definition of a p-n junction, it is probably fair to say that no one has yet made a true organic p-n junction.

Exciton Transport

Exciton transport lengths, L_{ex}, and the dissociation rate of excitons at interfaces, S (in units of cm/s), are clearly important parameters for understanding excitonic OPV cells. Unfortunately, L_{ex} and S are inextricably linked in most measurements and cannot be measured independently in a single experiment. Kenkre, Parris and Schmidt (23), in a classic study published in 1985, reviewed the literature on exciton transport length measurements from the preceeding 25 years and concluded that most of the results were incorrect. The problem was the almost universal assumption that S was effectively infinite (usually expressed as a boundary condition on the differential equation that set the exciton concentration near the quencher to be zero). Thus the measurements of L_{ex} were, in fact, just lower limits to the actual values and may have underestimated them by an order of magnitude or more. In fact, Kenkre, et al. concluded that the measurements in almost all previous studies had been limited by S, not by L_{ex} as assumed, and thus the actual values of L_{ex} were still unknown. It is only by finding and employing kinetically fast quenchers that L_{ex} can be accurately measured. Unfortunately, even now, many purported measurements of L_{ex} still rely on the unsupported assumption that $S = \infty$; thus many of the values of L_{ex} reported in the literature, now as then, are questionable.

We studied exciton transport in polycrystalline films of perylene-*bis*-phenethylimide, PPEI (24). Ten different quenching surfaces were investigated. Quenching at oxide semiconductors like ITO, SnO_2 or TiO_2 was uniformly slow ($S \sim 10^3$ cm/s); most excitons were reflected from these surfaces without quenching. The fastest quencher investigated was poly(3-methylthiophene) where S $\sim 10^6$ cm/s. Using the fastest quencher, a value of $L_{ex} \approx 1.8$ µm was measured whereas, if we had made the common, and commonly unsupported, assumption that a SnO_2 electrode is an efficient quencher, we would have concluded that $L_{ex} \approx 0.07$ µm. It is clearly important to know if a solar cell is being inefficient because most of the excitons are not reaching the interface, or because most of the excitons are being reflected from the interface without dissociating. But the information needed to make this determination is not yet available for many materials. If L_{ex} is actually shorter than the optical absorption length, $1/\alpha$, then the only way that all excitons can reach the interface in an optically thick cell is to structure the interface.

Structured Interfaces

OPV cells often work best when the organic film is quite thin. There are several factors that may contribute to this result including the high electrical resistance of most organic semiconductors, their low carrier mobilities, limited exciton transport lengths and dissociation rates, etc. Also, the equilibrium electric field is higher for thinner films. Dye-sensitized solar cells embody a "thin film" OPV cell taken to its logical extreme: the organic film is less than a single monolayer adsorbed to a substrate like TiO_2. Grätzel's key idea was to make the substrate so highly structured that even this ultrathin organic film would be optically thick (2). In DSSCs, the required exciton and charge carrier transport lengths through the organic film are effectively zero, thus eliminating two of the major limitations of organic materials. Of course, other limitations appear in such a device, but at present the DSSC is by far the most efficient OPV cell (~10%).

With the success of the DSSC, several groups started structuring the interface of other types of OPV cells leading to marked improvements in some polymer-based cells (6, 7, 9). The highest solar efficiency achieved so far in polymer cells with a structured interface is about 2.5% (9); about 1.4% has been achieved with a planar interface (and different polymers) (8). So far, little work has been reported on structuring the interface of molecular semiconductor cells, although a similar effect has been achieved with a light trapping device and a 10 nm thick organic layer resulting in an efficiency of 2.4% (5). An efficiency of 4.5% has been reported in single crystal pentacene doped with bromine (3).

All of these records have been reported within the last year, demonstrating the remarkable vitality of the field.

There are two ways to structure the interface of OPV cells, both of which can be illustrated by Figure 4.

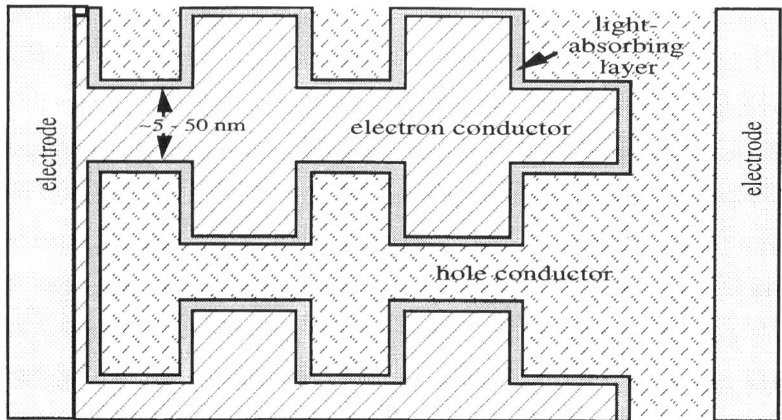

Figure 4. *A schematic representation of an OPV cell with a structured interface. The electron conducting and hole conducting phases are present in all schemes; they are ideally bicontinuous and each contacts only one electrode. The light absorbing layer is present only in some schemes, such as DSSCs, otherwise the light is absorbed by the electron conducting and/or hole conducting phases.*

The first way to structure an interface, of which the DSSC is an example, involves structuring the substrate and covering it with a thin layer of organic semiconductor or sensitizing dye. Electrical contact between the structured interface and the opposite electrode requires a third phase, shown as a hole conductor in Figure 4. The second way, of which polymer/polymer blends and blends of fullerenes or quantum dots in a conducting polymer are examples, is to use planar substrates and structure the interface between the two photoactive materials. In this case the thin film labeled "light absorbing layer" in Figure 4 is unnecessary because the electron conductor and/or the hole conductor absorbs the incident light. These two types of structuring have somewhat different characteristics that are described in bullet form below.

As the area of the interface increases (structured substrates) ...
* The thickness of the light absorbing layer decreases
* More excitons successfully reach the interface
* Charge transfer through the organic film improves

* The interfacial current density, and the harmful electric field it causes, decrease
* Interfacial recombination occurs over a greater area
* Pinhole density, and the risk of electrical shorts, increases
* Cell gets thicker, transport through "bulk" takes longer
* Making electrical contact across the whole interface and with the respective electrodes becomes more difficult
* It gets harder to make a solid state cell

As the area of the interface increases (structured organic films) ...
* The thickness of the light absorbing layer does not change
* More excitons successfully reach the interface
* Charge transfer through the organic film becomes slower
* The interfacial current density, and the harmful electric field it causes, decrease
* Interfacial recombination occurs over a greater area
* Not all pieces of a phase may be connected; need to be above the "percolation threshold"
* It becomes more difficult to ensure that each phase is contacted only by "its" electrode
* Ability to make a solid state cell is unaffected

The major advantages of structuring the interface, by either technique, are that more excitons will reach the interface and be able to dissociate, and that the deleterious photo generated electric field at the interface (discussed earlier) will diminish as the charge carriers are spread over a larger area. Each type of structuring also has specific advantages and disadvantages compared to the other: For example, charge transport through the organic film is more rapid for structured substrates because the film is thinner; but for structured films on planar substrates, charge transport can be slower because the carriers must traverse a more convoluted path, sometimes through narrow constrictions. Since charge transporrt through organic semiconductors is often slow, this could be a substantial disadvantage to structuring the film rather than the substrate. (With structured substrates, both electron conductor and hole conductor are usually much more conductive than the organic film.) A further disadvantage of structuring the organic film on planar substrates is the possibility of creating electrically isolated pieces of one phase when the two phases are mixed together, for example, isolated clusters of fullerenes or quantum dots may be created when mixing them into a conducting polymer. Charge carriers generated in these clusters can only recombine.

On the other hand, a substantial advantage of structuring the organic film over structuring the substrate is the ease of making an all solid state device, still a real challenge with highly structured substrates. Also, it is generally easier, and therefore potentially less expensive, to make cells with structured films rather than with structured substrates. Finally, cells made with structured films have only one interface where recombination is expected to occur while those made with structured substrates have two: the organic/electron conductor interface and the organic/hole conductor interface (Figure 4). This may result in lower photovoltages for the latter. This analysis notwithstanding, much remains to be learned about both of these configurations. Except for the DSSC, little has been reported yet about the use of structured substrates in OPV cells.

Summary and Conclusions

A brief overview was given of the physical principles by which solar cells function, with an emphasis on the role played by the chemical potential gradient. A PV effect can occur without band bending and, therefore, the measured photovoltage is not necessarily an indication of the band bending. Interfacial exciton dissociation plays a major role in many OPV cells and the consequences of this mechanism have not been clearly addressed before. Charge carrier generation and recombination in these cells occurs almost entirely at the interface; thus interfacial properties can assume greater importance than the properties of the bulk organic semiconductor. The photogenerated interfacial electric field caused by exciton dissociation opposes charge separation; at high light intensities, this field may overwhelm the equilibrium built-in potential. There is, at present, a regrettable lack of exciton transport measurements that take into account the finite rate of exciton quenching; most transport lengths quoted in the literature are merely lower limits to the actual values. The two techniques for structuring the exciton dissociating interface were compared and some of the advantages and disadvantages of each were discussed.

Currently, some of the highest efficiency OPV cells throw away the organic's major advantage over conventional solar cells – their simplicity and low cost. Its always useful to probe the efficiency limitations of a new type of PV cell, but ultimately, if OPV is to become a viable solar cell technology, it must be based on low cost, stable materials and simple, inexpensive processing. The remarkable progress in this field over the last few years suggests a promising future.

Acknowledgments

I am grateful to the U. S. Department of Energy, Office of Science, Division of Basic Energy Sciences, Chemical Sciences Division for supporting this research.

References

(1) O'Regan, B.; Grätzel, M. *Nature* **1991**, *353*, 737-740.

(2) Hagfeldt, A.; Grätzel, M. *Acc. Chem. Res.* **2000**, *33*, 269-277.

(3) Schön, J. H.; Kloc, C.; Batlogg, B. *Appl. Phys. Lett.* **2000**, *77*, 2473-2475.

(4) Schön, J. H.; Kloc, C.; Bucher, E.; Batlogg, B. *Nature* **2000**, *403*, 408-410.

(5) Peumans, P.; Bulovic, V.; Forrest, S. R. *Appl. Phys. Lett.* **2000**, *76*, 2650-2652.

(6) Yu, G.; Gao, J.; Hummelen, J. C.; Wudl, F.; Heeger, A. J. *Science* **1995**, *270*, 1789-1791.

(7) Granström, M.; Petritsch, K.; Arias, A. C.; Lux, A.; Andersson, M. R.; Friend, R. H. *Nature* **1998**, *395*, 257-260.

(8) Jenekhe, S. A.; Yi, S. *Appl. Phys. Lett.* **2000**, *77*, 2635-2637.

(9) Shaheen, S. E.; Brabec, C. J.; Sariciftci, N. S.; Padinger, F.; Fromherz, T. *Appl. Phys. Lett.* **2001**, *78*, 841-843.

(10) Fahrenbruch, A. L.; Bube, R. H. *Fundamentals of Solar Cells. Photovoltaic Solar Energy Conversion*; Academic Press: New York, 1983.

(11) Simon, J.; Andre, J.-J. *Molecular Semiconductors*; Springer Verlag: Berlin, 1985.

(12) Huang, S. Y.; Schlichthörl, G.; Nozik, A. J.; Grätzel, M.; Frank, A. J. *J. Phys. Chem. B* **1997**, *101*, 2576-2582.

(13) Zhang, Y.; Mascarenhas, A.; Deb, S. *J. Appl. Phys.* **1998**, *84*, 3966-3971.

(14) Pope, M.; Swenberg, C. E. *Electronic Processes in Organic Crystals and Polymers, 2nd Ed.*; Oxford University Press: New York, 1999.

(15) Gregg, B. A.; Fox, M. A.; Bard, A. J. *J. Phys. Chem.* **1990**, *94*, 1586-1598.

(16) Gregg, B. A.; Kim, Y. I. *J. Phys. Chem.* **1994**, *98*, 2412-2417.

(17) Gregg, B. A. *Appl. Phys. Lett.* **1995**, *67*, 1271-1273.

(18) Halls, J. J. M.; Walsh, C. A.; Greenham, N. C.; Marseglia, E. A.; Friend, R. H.; Moratti, S. C.; Holmes, A. B. *Nature* **1995**, *376*, 498-500.

(19) Haugeneder, A.; Neges, M.; Kallinger, C.; Spirkl, W.; Lemmer, U.; Feldman, J.; Scherf, U.; Harth, E.; Gügel, A.; Müllen, K. *Phys. Rev. B* **1999**, *59*, 15346-15351.

(20) Huynh, W. U.; Peng, X.; Alivisatos, A. P. *Adv. Mat.* **1999**, *11*, 923-927.

(21) Arango, A. C.; Johnson, L. R.; Bliznyuk, V. N.; Schlesinger, Z.; Carter, S. A.; Hörhold, H.-H. *Adv. Mater.* **2000**, *12*, 1689-1692.

(22) Pichot, F.; Gregg, B. A. *J. Phys. Chem. B* **2000**, *104*, 6-10.

(23) Kenkre, V. M.; Parris, P. E.; Schmidt, D. *Phys. Rev. B* **1985**, *32*, 4946-4955.

(24) Gregg, B. A.; Sprague, J.; Peterson, M. *J. Phys. Chem. B* **1997**, *101*, 5362 - 5369. The value of L_{ex} quoted in the present work is somewhat shorter than in the referenced paper due to a revised value of the absorption coefficient.

Indexes

Author Index

Subject Index

P

Papers, molecular electronics, 4–7

Passive electrode, indium–tin oxide (ITO), 134

Permeability, double-layer capacitance measurements, 46

Perylene-*bis*-phenethylimide (PPEI), exciton transport, 253

Phenomenology, understanding electron transport, 18

Photoinduced electron transfer, solution-based systems, 19

Photoluminescence, tetrakis(4-(4'-(3",5"-dihexyloxystyryl)styryl)stilbenyl)methane, 189–190

Photovoltaics
  advantages of structuring interface, 255
  conventional solar cells, 247–248
  current density equation, 244, 246
  dye-sensitized solar cells, 253–256
  electrochemical potential of electrons and holes, 246
  excitonic solar cells, 248–252
  exciton transport, 252–253
  Fermi level, 246
  fluxes and forces, 244–247
  flux of electrons, 245
  indium–tin oxide (ITO), 134
  interest in non-conventional solar cells, 243–244
  open circuit photovoltage, 245–246
  organic, 253–254
  quasi Fermi level for electrons, 246
  schematic band diagram of conventional p-n homojunction solar cell, 245*f*
  schematic of organic, cell with structured interface, 254*f*
  structured interfaces, 253–256
  structuring types, 254–255

  See also Excitonic solar cells

Phthalocyanines
  attenuated-total-reflectance spectroelectrochemistry, 147, 150–151
  spectroelectrochemical studies, 135
  See also Indium–tin oxide (ITO)

Piranha treatment, indium–tin oxide (ITO) surface, 139, 146

Plastic substrates. *See* Dye-sensitized solar cells

Polymerase chain reaction (PCR), background of molecular electronics, 4

Pore size distribution, compressed TiO_2 powder films, 126

Porosity, compressed TiO_2 powder films, 126, 127*f*

Porphyrins. *See* "Molecular-wire" linked porphyrins on gold

Probe lithography, techniques, 10–11

Proteins
  controlling electron flow, 36–37
  See also Electron transfer proteins

Q

Quantum chemistry, origin of decrease in electronic coupling, 70–71

Quantum-dot cellular automata (QCA)
  calculated and experimental geometries for two-dot and four-dot molecular QCA cells, 237*f*
  computational advantages, 238
  elucidating geometric and electronic structure, 238
  geometry optimizations, 238
  information processing, 235–238
  quantum states encoding information, 236*f*

Quasi Fermi level, electrons, 246

8-Quinolinolato. *See* Electroluminescent materials